BIOENERGY

VISION FOR THE NEW MILLENNIUM

Editors

R. RAMAMURTHI

Sri Venkateswara University, Tirupati, India

SATISH KASTURY

Florida Department of Environment, Tallahassee, Florida, USA

WAYNE H. SMITH

University of Florida, Gainesville, Florida, USA

Enfield (NH), USA Plymouth, UK

SCIENCE PUBLISHERS, INC.
Post Office Box 699
Enfield, New Hampshire 03748
United States of America

Internet site: *http://www.scipub.net*

sales@scipub.net (marketing department)
editor@scipub.net (editorial department)
info@scipub.net (for all other enquiries)

Library of Congress Cataloging-in-Publication Data

Bioenergy : vision for the new millennium/editors, R. Ramamurthi, Satish Kastury, Wayne H. Smith.
p. cm.
Includes bibliographical references and index.
ISBN 1-57808-141-6
1. Biomass energy. I. Ramamurthi, R. II. Kastury, Satish.
III. Smith, Wayne H.

TP339.B54 2000
333.95'39--dc21

00-046316

ISBN 1-57808-141-6

Published by Science Publishers, Inc., Enfield, NH, USA
Printed in India

Preface

Energy is one of the basic requirements that influence and limit our standard of living and technological progress. The way energy is produced and consumed affects the environment and therefore is a key issue in sustainable development. The continuous demand for energy is resulting in depletion of non-renewable sources and an increase in environmental problems. As a result, there is a growing recognition of need for development of eco-friendly technologies for sustainable energy production and ameliorate pollutants. This urge for sustainable energy production is to the point that it is believed preserving the natural ecosystem will emerge as the central organizing principle of twenty-first century.

The increase in global industrialization for the economic development and implementation of modern technologies in agriculture to produce higher yields to meet the food needs of the world has resulted in health, social and environmental implications. To evaluate the relevant problems as well as sustainable alternative technologies for energy production, an Indo-US workshop on "Eco-Friendly Technologies for Biomass Conversion to Energy and Industrial chemicals" was held in Tirupati, India in September, 1996. This workshop was organized by the Sri Venkateswara University, Tirupati, India; University of Florida, Gainesville, Florida, USA and Florida Atlantic University, Boca Raton, Florida, USA, which has a sustained interest in and active programmes addressing biotechnology for clean environment and energy.

The workshop was held at a very opportune time when there was a widespread concern for the state of global environment and its rapid degradation and also the availability of sufficient supplies of energy to sustain development. The workshop addressed the issues related to the use of innovative, relevant, appropriate and eco-friendly technologies for energy and the environment. The latest technologies developed for generating energy using non-conventional methods, and available technologies developed for generating energy using non-conventional methods, and available technology for management of the environment and remediation, were discussed on the technical sessions. The papers given at this workshop are presented in this proceeding. These papers are reproduced in full with no technical modification by the universities. The views expressed are those of the authors.

R. Ramamurthi
Satish Kastury
Wayne H. Smith

Contents

Preface iii
List of Contributors vii

1. Bioenergy 1
R. Ramamurthi and G. Bali

2. The Industry Perspective on Eco-friendly Technology for Biomass Conversion into Energy 11
Richard Schroeder

3. Specific Issues Relating to Bagasse Usage 21
W.H. Smith

4. Effective Utilization of Bagasse in Cogeneration of Power in a Sugar Plant 27
D.C. Raju and G. Prabhakar

5. Biomass Energy Plants 35
Peter C. Rosendahl and D. Carson

6. Thermal Conversion of Biomass 41
Alex Green, Mauricio Zanardi and Sergio Peres

7. Woody Biomass Production 63
Donald L. Rockwood

8. Woody Biomass Production, Species, Land Availability, Policy Issues, Scientific and Technology Needs—Assessment 73
P.M. Swamy and C.V. Naidu

9. Value-added Chemicals from the Byproducts of Sugar Agro Industry 85
G. Prabhakar and D.C. Raju

10. Composting: An Eco-friendly Technology for Waste Management 91
Aziz Shiralipour

11. Eco-friendly Technologies for Environmental Remediation/Management 99
S. Kastury

12. Production of Ethanol from Lignocellulosic Materials Using *Clostridium thermocellum*—A Critical Review 107
G. Seenayya, Gopal Reddy, M. Sai Ram, M.V. Swamy and K. Sudha Rani

13. Cellulose Conversion to Ethanol by a Mesophilic Cellulolytic Bacterium 117
T. Vijaya and D.V. Dev

14. Lending in the Biomass Energy Sector 123
V. Bakthavatsalam

List of Contributors

Bakthavatsalam, V. Indian Renewable Energy Development Agency Limited, Habitat Centre Complex, Lodi Road, New Delhi 110003, India.

Carson, Donald. Flo-Sun Sugar Company, Palm Beach, Florida, USA.

Dev, D.V. Science and Biotechnology College, Dr. Babasaheb Ambedkar Maratwada University, Nanded, India.

Geeta, Bali, Department of Zoology, Bangalore University, Bangalore, India.

Gopal Reddy, M. Department of Microbiology, Osmania University, Hyderabad, India.

Green, Alex. Clean Combustion Technology Laboratory (CCTL) and Department of Mechanical Engineering, University of Florida, Gainesville, Florida, USA.

Kastury, S. Florida Department of Environmental Protection, Tallahassee, Florida, USA.

Naidu, C.V. Biotechnology Center for Tree Imrovement, Tirupati, India.

Peres, Sergio. Department of Mechanical Engineering, Combustion Engineer from CHESF, Professor of the University of Pemmbuco, Brazil.

Prabhakar, G. Department of Chemical Engineering, College of Engineering, Sri Venkateswara University, Tirupati, India.

Raju, D.C. Department of Chemical Engineering, College of Engineering, Sri Venkateswara University, Tirupati, India.

Ramamurthi, R. Department of Zoology, Sri Venkateswara University, Tirupati, India.

Rockwood, Donald. School of Forest Resources and Conservation, University of Florida, Gainesville, Florida, USA.

Rosendahl Peter, C. Flo-Sun Sugar Company, Palm Beach, Florida, USA.

Sairam, M. Department of Microbiology, Osmania University, Hyderabad, India.

Seenayya, G. Department of Microbiology, Osmania University, Hyderabad, India.

Schroeder, Richard. Kenetech Resources Recovery, Gainesville, Florida, USA.

Shiralipour, Aziz. Center for Biomass programs, University of Florida, Gainesville, Florida, USA.

Smith, W.H. Center for Biomass programs, University of Florida, Gainesville, Florida, USA.

Sudha Rani, K. Department of Microbiology, Osmania University, Hyderabad, India.

Swamy, M.V. Department of Microbiology, Osmania University, Tirupati, India.

Swamy, P.M. Department of Botany, Sri Venkateswara University, Tirupati, India.

Vijaya, T. Biotechnology Centre for Tree Impovement, Tirupati, India.

Zenardi, Mauricio. Energy Department, Sao Paulo State University, UNESP-Campus of Guaratingueta, Brazil.

1

Bioenergy

R. Ramamurthi and Geetha Bali

INTRODUCTION

On the occasion of the United Nations Conference on New and Renewable Sources of Energy at Nairobi in August 1981, the Food and Agricultural Organization (FAO) was anxious to draw attention to the global energy crisis, that of fuel wood, which affects the daily energy supplies of great many rural people in the Third World. There is no need to overemphasize the problem of utilization of bioenergy supplies for people who have little or no access to any other sources of energy. What is probably shown in a new and striking way is the dimension of the problems, the order of magnitude of the populations affected and size of the deficits, and also the accelerating degradation of the situations identified. Therefore, it seems clear, in fact, that the deficits are expanding and increasing rapidly under the combined effects of population growth, deterioration of natural resources and the absence of possibilities for replacement by other sources of energy. Unless there is a radical change in present trends, the people affected by energy shortage due to the scarcity of bioresources would have more than doubled by the year 2000, the deficit might correspond to half their energy needs. Such an energy gap can hardly be accepted sociologically, economically or politically, concerning as it does the most elementary needs of the people. To this is added the fact that the great majority of the people concerned are among the poorest. It can thus be imagined what a burden the necessity to organize emergency actions to guarantee a minimum energy supply and distribute it to the people scattered over vast areas would represent for developing economics. But together with energy gap, and even more serious, would be the often irreparable damage to the fragile ecology.

Bioenergy is more relevant to developing countries which have huge population problem when it comes to the question of choosing a renewable source. While the bioenergy production technologies are global the relative emphasis of its potential might vary and truly country or region specific. Bioenergy will definitely be one of the many simultaneous options for relieving the energy crunch in many countries.

The major contributions on bioenergy in recent years have been extensively reviewed (Barnard, 1990; Betters *et al.*, 1991; Hall *et al.*, 1991; Ravindranath and Gadgil, 1991; Beyea *et al.*, 1992; Williams and Larson, 1993; Ahmed, 1994; Ravindranath *et al.*, 1994; Ravindranath and Hall, 1995; Björjesson and Gustavsson, 1996).

Indian Situation

The General President for the 1980 session of the Indian Science Congress, Dr A.K. Saha gave a detailed analysis of energy strategies in India. I shall discuss the present energy sources with particular reference to bioenergy in this country.

India has today a very low per capita consumption of energy, about one-tenth of the world average. A large part (45%) is non-commercial consisting of fuel wood, animal and agricultural wastes, which has gone up significantly over the past decades, seriously affecting the environment, in terms of deforestation, erosion, loss of soil fertility, etc. 50 % of the total energy consumption is in the household sector.

Bioenergy (non-commercial) typically rural in nature involves the use of fuel wood, agricultural residues, animal manure, human and animal power and no significant relief has been achieved during 7th and 8th Plans except the availability of kerosene and diesel oil.

Global Status

Biomass fuels had been a very negligible factor in industrialized countries till 1973, after which they have been receiving more attention. These fuels are essential in third world countries mainly for domestic purposes. The most important source is still firewood (World Resources Institute, Tropical Forests—A call for action, Washington DC, 1985). More than 100 million people use fuel wood below the minimum requirement and by 2000 AD, 3 billion people will either be unable to obtain their minimum energy needs or be forced to use wood faster than its regeneration (Year Book of Forest Products, FAO, Rome, 1985). Crop residues and dung in addition provide almost 90% of the energy in rural areas of North India, Bangladesh and China (Barnard and Kristoferson, 1985). Agricultural residues are used as fuel in the third world (International Institute for Environment and Development, Earth Scan, Lond., p. 12).

The International Institute of Applied Systems Analysis (IIASA) and the World Energy Conference (1983, WEC) project large increases in energy consumption by 2030 AD based on 1980s financial and institutional trends. An alternative study, using a wide range of energy mixes and taking into account more efficient use, gives a significantly lower projection (half and less) for 2020 AD as compared to earlier estimates (Goldenberg *et al.*, 1985).

I shall now briefly enumerate the various types of bioenergy sources and their production/utilization with particular reference to India, which may have a bearing on energy situations in other developing countries.

RENEWABLE ENERGY

Animal Energy

Animal energy accounts for over half the energy consumed in the agricultural sector. India has about 80 million work animals, 70 million bullocks, 8 million buffaloes and one million each of horses and camels, in addition to donkeys, elephants and yaks. Taking the generation of horsepower as 0.5 per animal, the installed capacity of animal labour force is 40,000 hp or about 30,000 MW, almost equal to the actual electric power generation in the country, at about one-third the cost of the latter. The animal drawn carts are about 15 million approximately. The National Commission on Agriculture (FAO) (1976) had estimated that the number of bullocks by 2000 AD will be maintained at the same level, through the contribution of animal power to total available farm power may fall to about 20%. The animal-drawn carts will maintain their use by small land holders (below 5 ha) in village lanes and for short distances. The states of Punjab, Haryana, Rajasthan, Gujarat, UP, Bihar and West Bengal had a total of 6.4 million animal carts of at least 3000 designs in 1978 (National Council of Applied Economy Research, New Delhi, Survey).

Biogas

The biogas programme in India started immediately after Independence (Wise, 1981). It is estimated that 1000–1050 MT of wet animal dung is available annually from 237 million cattle. At 66% of the collection rate as much as 22,425 million m^3 of gas can be produced through biogas plants. It can replace 13,904 million litres of kerosene per year. The slurry left over can produce 206 MT of organic manure annually equivalent to 1.4 MT nitrogen. 1.3 MT phosphate and 0.9 MT potash as fertilizers. It also enriches the soil by improving soil fertility and soil structure leading to an overall increase of the agricultural production by about 30%.

Approximately 190,000 biogas plants were installed in 1985-86, the total number in the country now is more than 610,000. 150 thousand such plants are estimated to save 6 MT of wood equivalent and give an annual return of Rs. 500 million ($ 11.16 million) (Dept. of Non-conventional Energy Sources Report, 1985-86). However, their functioning is seldom at an optimal level, being greatly affected by size of the plant and socio-economic conditions.

Biogas technology is yet to be fully standardized. The variable factors include dung quantity and availability, gas yield, calorie value and appliance efficiency. A rural household requires 1.7 to 1.29 m^3 per day of gas equivalent to 14.3 bovines per households. Based on the ownership pattern of bovines (30–80%), the number of rural households (out of a total of 121 M) expected to use biogas exclusively for cooking is from 16.03 M to 22.10 M depending upon the level of gas yield.

Two major research needs are restricted use of water and more effective strains of methane-generating bacteria to operate at a wide range of temperature. The addition of other components, like 40% human excreta to the manure along with stalks and grass and 50% of water, is used successfully in China. Pig manure in India is available only in North-East from 6.5 M pigs. Agricultural residues (100 MT) are also available. Thus there is a need to develop multifeed stock biogas technology.

Biogas can be generated also from the sludge obtained after primary treatment of raw sewage, as in operation at Okhla, India. The remaining matter is an excellent manure, while the supernatant can be used for fishery and algal growth.

In 2000 AD, at the present rate, the country would need 333 MT of dungcake to be obtained as decentralized energy system (BESI Newsletter, 1986).

Biomass is a general term for all materials whose origin directly or indirectly can be traced to photosynthesis. It includes all new plant growth, residues and wastes, as also biodegradable organic effluents from industries. Biomass can be produced from energy forests, hydrocarbon plants and oils, etc.

Wood is the principal source of fuel (68.5%) in rural India, followed by oil products (16.9%), animal dung (8.3%), coal (2.3%) and others (3.4%). About 1000 ha of land is estimated to provide enough wood, from suitable plantation of fast-growing trees, to give about 3 MW of power, besides fuelwood and charcoal for 125 to 150 families. The different proposals for other uses include: alcohol from plant wastes with high percentage of sugar, briquetting of organic residues; charcoal, petro-crops for hydrocarbons; self-reliant energy system in agro-industries; use of aquatic and coastal marine biomass; energy from urban wastes.

In the Seventh and Eighth Five Year Plans of India production and conversion to conveniently transportable and utilizable form of biomass were emphasized. Biomass Research Centres were proposed to perform surveys of indigenous resources, genetic, physiological and nutritional aspects of crop plants, training of farmers and dissemination of improved methods of using biomass.

The positive aspects of firewood use are its renewable nature and role in conservation of soil water and biological resources. The negative aspect is that in rural areas, the large gap between demand and supply results in decimation of rarest cover with consequent environmental degradation.

The firewood demand will be of the order of 350–375 MT by 2000 AD against the 1996 level of 120–130 MT combined with part of the dung of biogas and/or as fertilizer. Approximately 247 MT will be used in villages and 69 MT in urban areas for mainly domestic purposes. Only 50 MT of fuelwood may become available from the natural forests.

There is an annual shortage of 48 MT of fuelwood in the country, which may reach 125 MT by the turn of the century. So for the next 20 years the average fuel/wood contribution from natural forests would be 0.75 T/ha/year and the remaining part will have to be met from plantations on nonagricultural land. Roughly 60 million hectares of such land are available, but it might be difficult to bring more than 50% under plantation.

The total current potential for energy from biomass in India, including firewood and all types of wastes is estimated to be 825×10^{12} Kcal annually.

Gasification of biomass may be used to harvest energy through thermochemical conversion to biogas, producer gas and pyrogas. The technology needs updating.

Improved Chullah Programme is based on the conservation of energy use. About 112 million homes use traditional stoves with very low efficiency (2–10%). Nearly 30 improved models of stove have been devised by DENS (Department of Non-conventional Energy Sources, Govt. of India), incorporating optimum size of burning chamber, air inlet, graters, baffles and dampers so as to raise thermal effciency to about 15–25%. These stoves can also use other fuels like coal cakes, cowdung and pellets. 500,000 improved stoves are expected to result in an annual saving of 420,000 tonnes of firewood.

Petroplants: Liquid hydrocarbons as substitute for liquid fuels may be obtained from species belonging Euphorbiaceae, Ascelepiadaceae, Apocynaceae, Utricaceae, Convolulaceae and Sapotaceae. Out of 386 species screened for hydrocarbon content, fifteen have shown promise. There are varying estimates of power generation from 1 ha of land through this method (1.875 KW to 3.75 KW, ABE 1985).

The use of biomass has not gained as much significance as expected due to (i) constraints on the actual land available for plantation, as against the estimated area of wasteland, (ii) the minimum inputs involved, in terms of rainfall, fertilizer and irrigation, for developing energy plantations in wasteland, (iii) the use of most of the 'agricultural wastes'—estimated to be 200 MT in India annually—for other purposes in rural and agro-industrial economy, (iv) lack of a good database, outlining the actual situation.

In meeting rural energy needs, decentralized systems, using different available renewable resources, present the best option. Rural energy centres have been started on a pilot scale in India and a few other developing countries to highlight combining of biogass, biomass and solar energy.

Ethanol from Biomass

The economic prospects for biomass conversion to ethanol is a complex issue which is tied not only to petroleum prices but also to the cost of producing fermentable sugars from biomass like sugarcane sledges and other agriculture wastes (Wyud *et al.*, 1991). Energy used in the transportation sector totalled 22 quads (1 quad = 10^{15} BTU) in 1989 and accounted for more than 60% of total petroleum consumption. Furthermore, the transportation sector, with its nearly total dependence on petroleum, has virtually no capacity to switch to other fuels in the event of disruption in the petroleum supply (Lyard *et al.*, 1991).

Air pollution is also an important factor motivating interest in alternative fuels at the global level; carbon dioxide is responsible for more than half of the projected anthropically-mediated climate change. Transportation fuels account for 27% of the 3.3 billion MT of carbon dioxide released annually in the USA from combustion of fossil fuels, vehicles account for 4.7% of total worldwide anthropic carbon dioxide emissions, with US vehicles responsible for 2.5% of total emissions (Energy Balances of OECD) and other countries, 1971 to 1990; IEA, 1990). The Environmental Protection Agency (EPA) has stated that significant long-term environment benefits are available from the use of ethanol in engines designed to take full advantage of their combustion properties. Currently, 100 areas in the United States exceeded the national ambient air quality standards (NAAQS for zone, effecting more than half the population of USA). Although, magnitude of anticipated improvements in the zone levels and air toxins is relatively small, 5 to 15%, they are significant because, zone reduction is so difficult to achieve.

Therefore, it is worth considering to evolve more efficient methods for the biochemical conversion of wood or agricultural residuals to ethanol. This appears to be one of the most efficient options for utilizing all of the biomass components results in a fuel which can be blended with gasoline (ethanol gasoline) : (15 to 85%) and then directly utilized in current transportation fuel infrastructure. In certain areas of the world, such as Brazil more than 98% of alcohol is used directly, while in North America, ethanol-gasoline blends predominate as a means of utilizing ethanol. Virtually, all the transportation ethanol utilized today has been derived from starch or sugar substrates, such as corn, sugarcane or tapioca. Bioconversion of lignocellulosic residues has yet to progress past the process developing stage (Lynd *et al.*, 1991).

Hydrogen Energy

Hydrogen has good prospects of being an energy source in the future for the following reasons:

1. Hydrogen when burned in air produces a very clean energy without pollution.
2. Hydrogen is by and far the most abundant element in the universe and can be produced from water which is inexhaustible.
3. Hydrogen is the lightest fuel and richest in energy per unit mass.
4. Hydrogen can be produced easily and one can use any source of energy to produce hydrogen.

Photoproduction of Hydrogen by Algae

In the biological solar energy conversion to hydrogen at least two systems can be distinguished:

1. Hydrogenase/containing eukaryotic green algae. Cell-free chloroplast systems supplemented with bacterial hydrogenase, and
2. Cyanobacteria (blue/green algae) which is the most efficient systems for biological solar energy conversion.

Hydrogen Production by Making Photosynthetic Organisms (Bacteria)

The approach is based on the belief that there are undiscovered or little worked out organisms with inherent characteristics of hydrogen production. The advantages of using marine photosynthetic organisms in large volume cultivation are many; the most obvious being the availability of salt-water. Salt-water abounds with many nutrients (including carbon dioxide, magnesium, sulphates, potassium, etc.) essential for growth of photosynthetic organisms in tropical and subtropical environment and diverse photosynthetic microoganisms can be grown throughout the year.

The department of DNES is currently operating 23 ongoing research and developmental projects in various reputed national institutes, universities and IITs. All these projects are aimed at production, storage and utilization of hydrogen.

INPUTS FROM SCIENCE TECHNOLOGY

The bioenergy programme has to be developed in consonance with ecological requirements and as part of rural and human resource development. The important points are:

1. Collection and classification of information on biomass through remote sensing technology, to build a data base of terrestial and aquatic ecosystems and the sequenced charges, including local consumption.

2. Extension of integrated systems through demonstration, audio-visual variation, training and education, accompanied by monitoring and feedback.
3. For biomass production, identification, physiological and pathological studies of fastgrowing multipurpose trees, suited to specific agroclimates, breeding, tissue culture and use of cell and recombinant DNA technology to evolve energy-rich firewood with properties tailored for resistance and adaptability.
4. For biomass conversion fluidized bed combustion technology; manufacture of paralytic carbon from woody plants as a substitute for fossil fuels; development of fast effective immobilized whole cell system; gasification; studies on availability, nature of biocrudes and development of agrotechnology of more than 400 petrocrops are identified.
5. For production technology, modification and utilization of derivatives of jojoba oil.
6. Identification of non-edible vegetable oils as substitute for diesel.
7. Utilization of algae and other acquatic biomass for biogas technology.
8. Diversification of substrates to utilize materials like tea, green leaves and bamboo waste.
9. Optimal utilization of biogas slurry to replace inorganic fertilizers.
10. Raising bacterial strains able to function on different substrates, at different temperatures and moisture content through mutation or recombinant DNA technology.
11. Designing smokeless stoves and kitchens with ventilation.

REFERENCES

Ahmed, K. (1994). Renewable energy technologies: A review of the status and costs of selected technologies, World Bank, Technical Paper, No. 240. World Bank, Washington.

Barnard, G.W. (1990). Use of agricultural residues. In *Bioenergy and the Environment* edited by J. Pasztor and L.A. Kristoferson 85–112. Westview Press, Boulder, Colorado.

Barnard, G.W., and Kristoferson, L.A. (1985). Simplified analysis of acid soluble lignin. *Journal of Chemistry Wood and Technology*, **11:** 465–477.

BESI Newsletter 2 April, 1986, p. 20.

Betters, D.R., Wright, L.L., and Couto, L. (1991). Short rotation woody crop plantations in Brazil and United States. *Biomass and Bioenergy*, **1:** 305–316.

Beyea, J., Cook, J., Hall, D.O., Socolow, R.H. and Williams, R.H. (1992). Toward ecological guidelines for large-scale biomass energy development. National Audubon Society, New York.

Börjesson, P., and Gustavsson, L. (1996). Regional production and utilization of biomass in Sweden. *Energy*, **20:** 1097–1113.

Goldenberg, J., Poutanen, K., Koiner, H. and Viikari, L. (1985). In *Annual Review of Energy*, **10:** 622.

Hall, D.O., Mynick H.E., and Williams, R.H. (1991). Cooling the green house with bioenergy. *Nature*, **353:** 11–12.

Lyard, J.D., Wyman, C.E. and Grohmann, K. (1991). Fermentation of wood sugars to ethyl alcohol. *Industrial and Engineering Chemistry*, **37(4):** 390–395.

Lynd, L.R., J.H. Cushman., R.J. Nichols and Wyman, C.E. (1991). Fuel ethanol from cellulosic biomass, *Science*, **251:** 1318–1323.

Ravindranath, N.H., and Gadgil, M. (1991). Natural resource management and sustainable agricultural development. CES Technical Report. Indian Institute of Sciences, Bangalore.

Ravindranath, N.H., and Hall, D.O. (1995). Biomass energy and environment, Oxford University Press, Oxford, New York, Tokyo 1995.

Ravindranath, N.H., Nambi, V.A., Indrani, S., and Vinod, V. R. (1994). *Prosopis julifora* and household energy. CES Technical Report, Indian Institute of Sciences, Bangalore, India.

Reddy, A.K.N., and Ravindranath, N.H. (1987). Barriers to improvements in energy efficiency. *Energy Policy*, **19:** 953–961.

Williams, R.H., and Larson, E.D. (1993). Advanced gasification based biomass power generation. In *Renewable Energy* edited by T.B. Johansson, H. Kelly, A.K.N. Reddy and R.H. Williams, 729–785. Island Press, Washington.

Wise D.L. (editor) (1981). Fuel Gas Production from Biomass, Vol. l, 158–161 pp.

Wyud, D.R.J., Meijlink, I.H.H.M., Vleesenbeed, R., Vander Lans, R.G.J.M. and Luyben, K.Ch.A.M. (1991). Effects of the aeration rate on the fermentation of glucose and xylose by *Picbia stipitis* (BS 5773). *Enzyme and Microbial Technology*, **12:** 20–23.

2

The Industry Perspective on Eco-friendly Technology for Biomass Conversion into Energy

Richard Schroeder

INTRODUCTION

Biomass production in the US has been developed because of identified needs, and a willingness to address these needs. Specifically, bioenergy industry has grown and developed based on:

1. Forest industries' need for residue disposal and increased utilization of tree resources.
2. A desire to find and use alternative energy sources to reduce dependence on fossil fuel.
3. Solid waste disposal issues relating to utilization of resources, recycling, and reduction of solid waste landfills. This applies to household refuse as well as agricultural residues, commercial waste, and unwanted waste from land use activities.

Recently the concerns with global environmental issues such as climate changes and carbon balance, and interest in the need for increased and optimum use of agricultural resources (land, machinery, and people) has led to more activity. However, in the US most of the actual industrial development has been because of the above listed reasons.

Kenetech Resource Recovery, Inc. (KRR) is not a research organization. The company generates profit and return on investment based on the implementation of emerging technologies in biomass recovery. This year (1996) KRR will handle over 500, 000 tons of biomass, in the form of waste wood and plant waste, and will produce marketable products from this material.

KRR's experience in harvesting, material handling, and development of actual biomass industries has provided researchers and developers with practical information. Some practical perspective is provided to help implement this industry in India and other countries.

HARVESTING AND MATERIAL HANDLING OF BIOMASS

KRR's experience covers the development of biomass delivery systems, the contracting for long-term supplies of biomass, and the actual harvesting and material handling of the material on a large scale. These components are essential to any successful development effort, and must be supported with actual operating experience and proven technology.

The personnel in KRR have been involved primarily from two perspectives–forestry energy harvesting and wood waste recycling. Each of these is discussed briefly below.

FORESTRY ENERGY HARVESTING

Until 1978 biomass as energy was primarily a by-product of conventional harvesting and milling operations. During the oil shortages of the late 1970s and early 1980s, efforts were increased in developing methods that specifically managed biomass in forests as a fuel source.

During this period whole tree chip harvesting was developed for noncommercial trees, that is those trees that had too little value as timber or paper products to be conventionally harvested, or those trees that needed to be removed to promote forest growth. These harvesters changed to

Figure 1. A typical whole-tree harvester developed in the 1970s. Models are manufactured with diesel engines generating from 200 to 700 horsepower.

accommodate different types of tree forms and shapes, including trees with more branching larger diameter trunks, smaller stems, stumps, tops, and branches. Figures 1 and 2 land illustrate some of the commercially available equipment developed for energy harvesting.

During this time the personnel at KRR were involved in the process. Logging studies were conducted with the Florida Division of Forestry, and yield tables were derived for energy harvesting. In 1984, KRR managers operated a whole tree chipping operation for fuel, utilizing inmate (prisoner) labor, conventional harvesting equipment and harvesting noncommercial species and overcrowded stands of slash pine (*Pinus elliotti*). The material was delivered to a central boiler where it was utilized for steam and serviced the energy needs of the adjacent prison.

Experiments were conducted in different equipment/manpower combinations. Harvesting was done in various methods, from using practically no equipment and all manpower and consisting of men with chain saws cutting and carrying trees to small portable chippers to complete machine-harvesting, using a mechanical feller-buncher (Figure 3), a log skidder (Figure 4), and a whole tree harvester as shown in Figure 1. The experiments were possible because the inmate labor was available at very low cost and the property and trees were owned by the prison system.

Figure 2. Later generation tree harvester, designed for handling larger, less uniform trees, tops and branches.

From these studies some observations were made about biomass harvesting:

1. Under the conditions in the study, even with no-cost labor, the totally manual system did not yield the least expensive fuel material. This was because the supervision, the small amount of capital, the hand tools, and the safety and transportation expense was producing a very low output. Under the conditions of the study, less than five tons weighing 2,000 pounds including moisture (green tons) were harvested per man per day.
2. However, the most mechanized was not the least expensive. This was because the education, skill level, and motivation of the personnel was insufficient to effectively utilize the capital cost of the equipment, equal to $ 60000 or more in 1996 costs.

Figure 3. Typical mechanical feller-buncher, designed to cut trees lift them and place them in piles for removal by a log skidder.

From an industry perspective, because of the high cost of labor in the US and the currently low cost of capital (i.e., low interest rates), there is little interest in labor-intensive harvesting development. There is experience with these systems, though, and the method may apply in areas where support for machinery (parts, fuel, service expertise and trained operators) is not fully developed. A profit-making company will analyze the difference; in some cases it is more profitable to rely on large quantities or labor than it is to build the required infrastructure.

Figure 4. Log skidders designed to drag trees from the forest to be loaded on trucks or to a tree harvester.

WOOD WASTE RECYCLING

In the 1980s another issue emerged in the US that facilitated the biomass industry-environmental concerns became the reason for developing biomass energy. The solid waste generated by people in the US was becoming a major issue, and policies to recycle and manage waste streams began to be developed. Concerns about the emissions from fossil power plants, and an unwillingness to invest in nuclear power forced utilities to search for alternative energies. Wind, solar, and biomass were some of the alternatives, and one of the first sources examined was wood and biomass generated from agricultural, forestry, and solid waste refuse collection.

The principal business of KRR is wood waste recycling. Waste biomass in the US is generated not only by the agricultural industries such as sugarcane bagasse, by-products of papermaking, and other crop residues. It is also generated by households and commercial establishments. In the state of Florida, an estimated 1,000 pounds of wood waste is generated per family household per year. This consists of yard waste, scrap wood from construction and tree debris from storms.

In addition, commercial wood such as railroad ties, pallets, crates, and scrap from wood-using industries such as cabinetry, housing construction, and tree debris from landclearing generate a large amount of material annually. It is this material and the household wood waste that KRR recycles, and an industry has developed for recovering and utilizing these materials.

As in the case of forestry, the technology for recovery, or harvesting, has been changing rapidly. Initially machinery was taken from the forest products industry—whole tree chippers, bark grinders, and similar equipment. Eventually, machinery was developed exclusively for wood waste, using both forestry and agricultural technology.

The principal method to process wood waste is with a tub grinder, a machine orginally designed to grind hay and agricultural residues. After ten years of development, the tub grinder has been modified and developed to process wood wastes varying in consistency from wet leaves to stumps five feet in diameter. Figure 5 illustrates a typical tub grinder.

KRR harvests biomass wood in the operation described above and delivers it to biomass users. In addition, we have been involved in the early development of dedicated feedstock supply systems utilizing specialized crops, through development of projects in Minnesota, Puerto Rico and the United Kingdom.

In these systems crops are handled using agricultural machines such as forage harvesters, hay balers, or cane harvesters. Depending upon the crop

Figure 5. A tub grinder designed to process wood waste. In the US these machines are operated to recycle wood into fuel or soil products; KRR operates 12 of these machines in Florida

and the desired level of processing, existing equipment may be modified to enable a smoother implementation process. However, in the US these systems are still in the development stage, and commercial development is limited to applications of particular research projects.

ENERGY INDUSTRY REQUIREMENTS FOR DEVELOPMENT

The successful development of a biomass industry in India or other countries, whether it is based on forestry, wood waste recycling, or dedicated crops, relies on several criteria. The obvious one is economic—the return on the investment must be sufficient to cover the risks associated with the endevour.

In the US virtually all of the large-scale energy projects are accomplished with debt financing. Usually 65 to 90% of the required funds for construction is obtained from financial institutions, most of which have little knowledge of biomass as an industry. Because of this, the developer must understand the sensitivities involved with third party financing, as well as operational and technological issues.

As an active commercial developer of biomass industries, each potential opportunity is assessed based on the following requirements:

1. *Is the potential energy user willing to commit to a long-term purchase of energy?*
   Financing a project requires the revenue to be a guaranteed quantity for a guaranteed period of time. Small projects (less than 500 kilowatt per hour electrical output (5 MWe) capacity) may be achievable with 10 year energy purchase agreements, larger projects need 15 or 20-year commitments.
2. *Is the potential energy user a financially strong entity?*
   This is a requirement related to the financing arrangements. If third party financing is required, then the energy user must meet the financial strength requirements of the investors.
3. *Can the required amount of biomass energy be produced and transported to the energy user?*
   This requires some preliminary engineering and study, and can be facilitated by government, academic or private entities that have interest in community development.
4. *Can the cost of delivering the biomass be accurately predicted for the life of the project?*
   In the US, many projects have encountered trouble when the political, regulatory, or social climate changes. Predictability is important for long-term projects such as biomass development.

5. *Is the difference between the current cost of energy and the proposed biomass alternative sufficient to provide returns proportional to the risk?*
   In most cases biomass is an alternative strategy for energy production, and is either replacing or supplementing other energy sources such as fossil fuel. KRR has vast experience in quantifying the cost to deliver biomass, and the cost of conventional fuels is well documented. This difference combined with the difference in converting the two forms of energy equals the energy savings, and this savings must be sufficient to both stimulate the energy user to implement the conversion and provide a return on the investment. Our experience is that if an energy user can save 10 to 20% of their existing energy cost, and the simple return on the investment is in the 10 to 20% range, then the project is realistic.
6. *Are the government, community, and citizens receptive to the project?*
   Millions of dollars in the US have been spent and lost attempting to develop projects that make economic sense but that the host communities do not want. The best approach to this problem is to include the community as a benefactor, through tax payments, employment, or other benefits.
7. *Can engineers, contractors, and equipment vendors be located that are available at the project location, and can warrant their performance in the design, construction, and operation of the project?*
   This is especially true of international projects. For the most part these contractors will travel anywhere in the US for a project, but there are fewer that are capable of performing services internationally.

There are many other issues, but those outlined are considered as 'requirements'. If during preliminary investigations any of these are found to be negative, then it becomes difficult to interest a developer to expend effort or find a financial institution to invest capital to build the project.

INDUSTRIAL AND INFRASTRUCTURE NEEDS

As a biomass supplier we are concerned with transportation and production issues. While there are other issues relating to energy transmission and distribution and plant permitting, these are not discussed here.

For purposes of this discussion it is assumed that a large scale development (25 MWe or greater) is being considered. Smaller scale operations require less infrastructure support because impact is minimal. Using a 25 MWe plant as an example, some estimations of infrastructure requirements can be made.

With conventional technology and operating at full capacity (8,000 hours per year) this plant will require 138,800 tons of biomass with all moisture removed (dry tons) of biomass or about 231,400 green tons (40%

moisture content), per year. Using this as a base point, each of the components of a delivery system can be analyzed and the infrastructure assessed.

Production

In the case of dedicated biomass crops, if high yielding crops of 20 dry tons per acre per year are used, then the project will require 6,940 acres. Is the land available, and what is the production capability? How close to the facility can the crops be placed? Are there sufficient sources of land cultivation equipment, labor, and supplies such as fertilizers, pesticides, and water to provide for this production?

In the case of residues, the questions are more related to storage. Is the material produced seasonally, and, if so, is there room to store material to make it available year-round? How close is the residue to the project? Is more than one type of fuel going to be required in the project?

Harvesting

There are two basic approaches to designing a harvesting system. The first approach is to maximize labor and reduce the investment capital required. This is used where proper support for large machinery may be difficult or where creating jobs is one of the highest priorities.

Productivity of these systems is difficult to characterize. In the studies at the Florida prison harvesting noncommercial trees with manual labor resulted in about two green tons per labourer per day. Based on that result, the 25 MWe plant will require about 316 people working every day of the year to supply the project.

On the other hand, if complete mechanization is desired, then the plant will require about three or four fully mechanized harvesting systems, consisting of over $2,000,000 in equipment purchase costs. Using these as guidelines, an assessment of the capability of the labor market, the equipment support system, and the available capital can be used to select the most appropriate system. In most cases the optimum system will be a combination of labor and mechanization.

A project such as this involves the shipment of over 925 green tons per day on a five-day-per-week basis. If vehicles can be used that contain 24 tons of green weight per trip, then the project requires 39 truck loads per day for a five day week. Characteristically, most biomass fuel deliveries take over 3 hours, including loading, unloading, and transportation both ways. Based on this the project will need about 13 trucks pulling three loads per day.

This analysis is only valid if large trailers can be accommodated on the roads. In the United Kingdom it was found that many roads were too narrow for this type of equipment, which is 104 inches wide and may hold 120 cubic yards of material.

This analysis provides planners and interested organizations with an idea of the required infrastructure to implement a biomass project. The roads, vehicle weights, speed limits, taxes, labor rates, and cultural issues such as working hours and work days, division of labor, and educational level of the labor are all additional considerations.

SUMMARY AND CONCLUSION

The US has seen the development of a biomass industry that has involved research, development, and implementation by private firms based on the opportunities presented for return on investment. Initially harvesting and delivery systems were derived from forestry applications; later recycling activities led to additional equipment for processing wood waste. The next stage of development will be dedicated crops, utilizing crop harvesting technology.

KRR has been involved in the growth and development of the biomass industry. The company has operated harvesting and delivery systems for biomass in the US, related to both forestry and recycling or solid waste applications. From their experience the requirements and infrastructure needs have been outlined which will apply to any large-scale commercial application of biomass energy.

Conversely, if all the requirements are met and there is need for additional energy, then biomass represents a viable alternative to other forms of energy. The environmental benefits are known, and many communities recognize the value in placing local resources to work for generating local energy needs instead of exporting currency to other areas.

Research and business leaders in India and other countries considering biomass development should study and utilize the experiences gained in the US. Not all efforts have been successful, but the efforts have yielded an industry that can help implement these projects in other places more quickly and successfully.

3

Specific Issues Relating to Bagasse Usage

W.H. Smith

INTRODUCTION

Plants through photosynthesis offer a means of capturing solar energy and storing it as chemical energy. Sugarcane belongs to a group of plants known as C-4 plants, that is most efficient in this process. These plants grow best at elevated temperatures and use more of the light absorbed than C-3 plants and they do not lose about 30% of the already fixed carbon through photorespiration. Taken together, these C-3 (mainly temperate) and C-4 (mainly tropical) plants can process enough solar energy to annually produce an estimated 220 billion dry tonnes of biomass equivalent to about 10 times the global energy used (Hall *et al.*, 1993).

Because of distribution, harvesting, user specification and demand characteristic, biomass provides only 15% of the world energy use. In developing countries, however, about 38% of the energy use is from biomass and in specific countries it can approach 95 per cent (Hall *et al.*, 1992).

Biomass has been harvested and used directly for energy since the use of fire became manageable. However, the 'gather and use' strategy is inadequate when work and services from energy went beyond subsistence. For many decades residues from cropping or processing enterprises proved adequate to meet these needs. In some cases these residues helped expand local processing capabilities and in some cases meet other needs. For example, wood wastes have long been used for drying and for providing other on site needs for energy in the forest products processing. In other cases, residues have helped meet off-site needs for energy. For example, surplus bagasse from the sugarcane industry in Hawaii formed the primary fuel for initiating a community electrical utility.

Bagasse from sugarcane made this possible because each ton resulted in about 250 to 280 kilograms of bagasse depending upon harvesting and processing practices. The sugar mills use bagasse for process energy. Worldwide these mills have been inefficient, operating largely as incinerators for waste disposal while yielding process heat. After environmental constraints became more rigorous and fossil fuels became more expensive, incentives were offered for considering other or more efficient ways to use the bagasse. Sugar mills in Florida have recently constructed efficient electrical generating plants to use bagasse. FLO-SUN ENERGY is also exploring various 'bridge fuels' for the non-grinding season.

ISSUES RELATING TO BAGASSE USAGE

Alternate Uses

Depending on process efficiency, surplus bagasse at a sugar mill can represent 8 to 15% of bagasse produced. Some bagasse has been used in limited ways as animal feed after various pretreatments. Because of advances in pulping processes, increasing quantities of bagasse are being used in newsprint manufacturing. For example, the Tamil Nadu Newsprint and Paper Project in India that developed a 300 tonnes per day bagasse newsprint mill (Atchison, 1992) is typical of several recent successes. These have led to an increase in the pulp capacity for using bagasse from 120,000 tonnes in 1950 to 2.6 million tonnes in 1992. Bagasse pulps are now used in practically all grades of paper including bag, wrapping, printing, writing, toilet tissue, toweling, corrugating, board stock, etc.

In recent years, considerable interest has focused on converting bagasse to ethanol, largely because of the need to extend or enhance transportation fuels. Such interest has been stimulated by genetic engineering advances that now make it possible to ferment diverse sugars (e.g., pentoses as well as hexoses). An example is the use of the organism resulting from the 5 millionth USA patent that covered a process for moving genes from an unrelated microbe (Ingram and Conway, 1988) to enhance alcohol fermentations. This process and the one developed by the National Renewable Energy Laboratory (NREL, USA) (Zhang *et al.*, 1995) has had limited commercial opportunity because of the need for low-cost methods to hydrolyze celluloses to sustain affordable sugar streams. Nevertheless, a recent analysis (Stricker *et al.*, 1995) by my colleagues working with private sector partners suggests the technology to be feasible for alcohol fermenting with electrical generation using residues and the waste heat used to provide process energy for the alcohol fermentation. Similarly, Wyman (1995) established a set of conditions and assumptions to show that economically affordable ethanol from lignocellulosic biomass is feasible using the NREL process.

Co-generation, especially when it is encouraged by incentives as in the USA is increasing the capture of bagasse for fuel. You will hear of the example in Florida where waste heat from electrical generation is meeting process energy needs for the sugar mill while producing about 40 MW of electricity at two different mills. Another example comes from Brazil (Guerra and Steger, 1988) where about 12% of the bagasse that is in surplus from a sucrose to ethanol facility is used to generate electricity. The "waste heat" is used to concentrate citrus juices and to dry citrus residues.

Many chemicals other than alcohol (e.g., furfural, ammonia via synthesis gas, and various other organic solvents) can be made from bagasse. However, the capital cost, product value and inexpensive alternative feedstocks have prevented their viability (Natham, 1978). ·

Scale Considerations/Consolidating Bagasse Supply

Capital cost is not highly variable at small scale with a 5 MW plant costing essentially the same as a 10 MW plant. But at some point, the cost per MW approaches a constant. A conventional biomass to energy plant in the USA typically costs from $1500–2000/KW. While there have been advances in the efficiences of conventional plants, there still exists economics of scale. Information available to us on sugar mills in India suggests that they are small and dispersed. Because of bulkiness there are limits to the distance that bagasse can be transported and still be an economically affordable fuel. Spatial dispersion and transportation optimization schemes are essential for consolidating the bagasse and defining the appropriate scale for power generation.

Available Technology

Recently emerging technologies relating to gasifiers and gas turbines show promise for reducing the capital cost per MW to the range of $500–600 per KW (Williams and Larson, 1996) and making smaller scale electrical generating plants feasible. The availability and adaptability of this technology needs to be explored and its fit to local Indian conditions determined. Similarly, recent and detailed analysis (Hatzis *et al.*, 1996) of new technology for ethanol from lignocellulosic biomass are encouraging and suggest that this technology deserves careful examination.

Sustainable Fuel Supplies

A problem with bagasse fueled power plants besides those noted above is the lack of a fuel supply when the cane grinding season has passed. In Florida FLO-SUN ENERGY relies on yard waste and other biomass residues as the "bridge fuel". I understand they are exploring tree plantations and salvage harvests of local invasive woody species. In the future, they may consider biogas produced from aquatic species grown to scrub nutrients from drainage waters of the Everglades Agricultural Area. The waste treatment/biogasification process has been tested at various sites including one operated by the University of Florida at the Walt Disney World Resort (Hayes *et al.*, 1987). Also, the University of Florida has researched the production of woody species and the tall grasses as biomass crops combined with bagasse could provide a sustainable fuel supply, especially if wastes can be recycled to maintain the fertility of the crop production media. In India both terrestrial and aquatic biomass species grown for fuel and/or other biomass resources need evaluation.

Environment Concerns

Developing the opportunity to use bagasse for fuel may create pressure to use other cane fractions. About 30% of a cane plant is tops and leaves. These are usually burned or otherwise left in the field to recycle their minerals and/or organics content. These materials should be removed, their nutrient content would have to be restored with proper fertilization and practices be implemented to prevent erosion.

Ash from cane constitutes about 6% of the dry weight (Nathan, 1978). This could vary depending upon the amount of soil attached to the plants and the soil growth medium. This ash could be recycled to the fields in various ways even in fertilizer mixers. As for stack emissions, there exists best available technology appropriate for each regulatory environment.

If there exists waste water streams that could be cleansed with aquatic plants, environmental enhancement could occur while providing a biogasification feedstock. There are also environmental concerns associated with transportation that have to be considered. These relate to traffic congestion, road degradation, noise and dust. These concerns have to be considered in any site specific analysis.

Financing

The viability of renewable energy systems, including those involving biomass have mainly been limited by economic considerations rather and technical feasibility. In fact most biomass systems are tied to some kind of economic incentive to bring them over the threshold for commercialization.These are often essential because conventional fossil fuels and domestic crops (especially in developed countries) have enjoyed subsidies for many years allowing technological advances and energy advocates to have subsidies (e.g., large annual research development and demonstration). Our challenge is to develop the best available technology scheme and subject it to rigorous economic analysis for financial experts from the private sector and those from agencies such as the USDOE/AID, the Indian Department of Non-conventional Fuels, the World Bank and other to carefully evaluate to gauge their investment prospects in financing a project for India.

SYSTEMS ANALYSIS

Developing sustainable electrical generation based on bagasse clearly depends upon proper resolution of several issues. These have to be cast in a relational framework subject to a system analysis. This framework will include models that will allow the simultaneous treatment of each issue for assessing system cost and output and their sensitivity to the several variables affecting the bagasse to energy process. Several models for such analysis have been developed at the University of Florida. We have advanced models for both a biogasification system (Mishoe, 1988, Legrand, 1993) and for a dedicated feedstock supply system for electric power and ethanol in a Central Florida location (Stricker *et al.*, 1995).

CONCLUSIONS

Sugarcane has proven to be one of the most productive crops grown woldwide. While 10 to 12% of the dry weight is sucrose and other potentially fermentable solids (Nathan, 1978), the resulting bagasse constitutes a useful resource for a number of products including fuel for electrical generation. However, sustain electrical supplies essential to local economic development, sugar mills should be show that significant surplus fuel results. Energy generation should use the most efficient conversion equipment to maximize power generation per tonne of biomass. To encourage regional economic development, electrical generation must be sustained year round and not just during the grinding season. Woody species, tall grasses or aquatic plants grown to renovate polluted water could meet this need.

All of these variables need to be cast in a system analysis to determine process sensitivities and optimum fits. The University of Florida and Sri Venkateswara University have experience in these areas and should work with the industries and agencies to explore the applicability of the new technologies to a local Indian situation. The FLO-SUN ENERGY model for using cane bagasse to generate electricity while capturing waste heat represents a functioning useful model to guide the discussions in this workshop and on other occasions.

REFERENCES

Atchison, J.E. (1992). Making the Right Choices for Successful Bagasse Newsprint Production, Part 1. TAPPI, December, p. 63–68.

Guerra, Jose Luiz and Steger Elizabeth (1988). Sugarcane bagasse: An alternative fuel in the Brazilian citrus industry. *Food Technology*, May **138–139:** 264–265.

Hall, D.O., J. Woods, and J.M.O. Surlock, 1992. Biomass Production and dates, appendix C. In *Photosynthesis and Production in a Changing Environment* edited by D.O. Hall *et al.*, Chapman and Hall, London.

Hall, D.O., F. Rosillo-Callie, Williams, R.H. and Woods, J. (1993). Biomass for Energy: Supply

Prospects. In *Renewable Energy* edited by T.B. Johansson, H. Kelly, A.K.N. Reddy, and R.H. Reddy, 593–651. Island Press, Washington, D.C.

Hatzis, Christos, Riley, Cynthia and Philippidis, George P. (1996). Detailed Material Balance and Ethanol Yield Calculations for the Biomass-to-Ethanol Conversion Process. *Applied Biochemistry and Biotechnology*. Vol. **57-58:** 443–459.

Hayes, T.D., Reddy, K.R., Chynoweth, D.P., Bilietina, R. (1987). Water Hyacinth Systems for Water Treatment. In *Aquatic Plants for Water Treatment and Resource Recovery*, edited by K.R. Reddy and Wayne H. Smith, Florida, Magnolia Press, Orlando, 121–139.

Ingram, L.O. and Conway T. (1988). Expression of different levels of ethanol genic enzymes from *Zymomonas mobilis* in recombinant strain of *Escharichia coli*. *Applied and Environmental Microbiology*, **54(2):** 397–404.

Legrand, R., (1993). Methane from Biomass System Analysis and CO_2 abstract. *Biomass and Bioenergy*. **5:** 301–317.

Mishoe, J.W., Boggess, W.G., Fluck, R.C., Kiker, C.F., Kirmse, D.F. (1984). BIOMET-Model of Biomass to Methane Production and Conversion System. Documentation and Status Report. For IFAS & GRI. University of Florida.

Nathan, R.A. (1978). Fuels from sugar crops. Technical Information Center, USDOE, Oak Ridge, Tennessee, p. 137.

Smith, W.H., Wilkie, A., Smith, P. (1990). Methane from Biomass and Waste Programs in the USA. Proceedings of the International Conference on Energy and Environment, King Mongkut's Institute of Technology, Bangkok, Thailand, p. 248–269.

Stricker, J.A., Rahmani, M., Hodges, A.W., Mishoe, J.W., Prine, G.M., Rockwood, D.L., Vincent, A. (1995). Economic development through biomass systems integration in central Florida. Final Report to the National Renewable Energy Laboratory. Golden, Co. 122 pp. plus appendices.

Williams, R.H. and Larson, E.C. (1996). Biomass gasifies gas turbine power generating technology. *Biomass and Bioenergy*. **10:** 149–166.

Wyman, C.E. (1995). Economic fundamentals of ethanol production from lignocellulosic biomass. In: *Enzymatic Degradation of Insoluble Carbohydrates*. ACS Symposium Series 618, Washington D.C., p. 272–290.

Zhang, M.C., Eddy, Deanda, K., Finkelstein, M. and Picataggio, S. (1995). Metabolic Engineering of a Pentose Metabolism Pathway in Ethandogenic Zymomones Mobils. *Science*, **267:** 240–243.

4

Effective Utilization of Bagasse in Cogeneration of Power in a Sugar Plant

D.C. Raju and G. Prabhakar

It is a matter of serious concern that the generation of electric energy in India is not keeping pace with rise in demand. The total installed capacity till 1993 was 71,300 MW against the projected demand of 103,000 MW by the end of the Eighth Five Year plan. Including capacity additions from private sector power projects, the total installed capacity would be 99,620 MW leaving a gap of 4,000 MW unfilled.

It is estimated that demand for electric power is increasing at a rate of 9% annually in India, higher than the projected annual rate of 2.8% worldwide. The present per capita consumption of electric power in India is low at 36 watt as compared to 120 watt worldwide and 2,900 watt in the United States. Unless urgent corrective measures are initiated, the situation could prove to be catastrophic.

Traditionally, all energy comes from hydel, thermal and nuclear power plants in India and this overdependence on conventional sources cannot continue forever. Due to vagaries of the monsoon and uneven levels of rainfall across the country, the prospect of hydel reservoirs continuing to serve as future energy sources is not encouraging, while depletion of fossil fuels is the main cause of worry in case of thermal power plants. Presently, about 40% of total energy generated in the world is through oil. British Petroleum estimated in 1989 that existing oil sources would all be depleted in about 44 years. Even though the nuclear option is not fully exploited till date, it cannot be the future source of power for the simple reason that risks and hazards associated with it are too serious to be ignored. Hence there is an immediate need for identifying alternate sources and technology of power generation acceptable to the vast population of the country.

Efforts are already under progress in India and several other countries to make use of non-conventional sources like solar, wind, tidal and biomass energy. Inspite of best efforts by the scientific community and the government, technology for the exploitation of solar, wind and tidal sources is in the preliminary stages of development and does not offer much scope for use in the near future. India, being an agriculture based country, is rich in natural fauna and flora and produces *enormous* amounts of biomass. As biomass is renewable, energy generation based on it will be sustainable and also environment friendly. Whatever carbon dioxide is utilized by the plant while growing is returned to the surrounding environment, when it is burnt in producing steam/energy. Biomass is the future source of energy for India.

Cogeneration is an extremely useful strategy of producing electric power and heat energy from a common fuel. Sugar industry, which produces considerable amounts of biomass, bagasse, as a low cost by-product during its normal operation of making sugar from sugarcane, is the ideal choice for adoption of cogeneration. Cogeneration in sugar mills enjoys several advantages—low capital investment, low payout time, low gestation period, low transmission and distribution losses, better utilization of bagasse and high profitability. Further, this method results in considerable savings in the conservation of fossil fuels.

In India, annual production of sugarcane is averaging around 200 million tonnes. This cane, when processed in a sugar mill, yields about 11 million tonnes of sugar and about 40 million tonnes of bagasse. Most of the Indian sugar mills are of 2500 tonnes crushing capacity per day and it is estimated that, with the bagasse they produce, about 11 MW of electric power can be generated. About 3 MW of power is used up by the sugar mill itself leaving 8 MW of power for export to other consumers. The consolidated figure of electric energy available from over 400 sugar mills situated in India is about 3200 MW per annum.

Even though the general principles of cogeneration hold good for all the sugar mills, the specific details vary from plant to plant. What is good for one plant may not suit other plants. To design an economically viable cogeneration system, process variables like steam pressure, steam requirement and working pressure and temperature of the boiler and system parameters like generator voltage and power requirement of the plant serve as the key inputs. Two broad schemes can be tried in case of existing plants opting for cogeneration.

1. If the boiler available in an existing factory produces such quantities of steam that is equal to the steam requirement of the plant, such a system is said to be balanced. Additional power generation will then be possible only with the replacement of the existing boiler by a high pressure boiler. The high pressure, high temperature steam thus obtained can be fed to a

backpressure turbogenerator, which produces electric energy. The exhaust of the turbine will be at conditions suitable to the sugar plant and is directly fed.

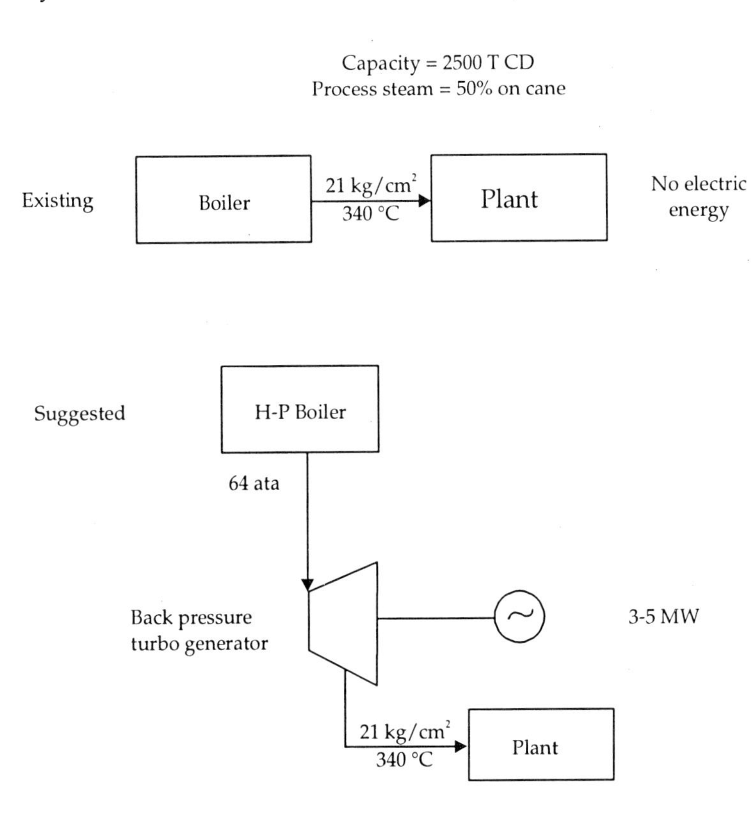

Scheme I

Here all the bagasse produced by the sugar mill may not be used. Some bagasse is left over and it can be stored for off season power generation.

2. If the boiler steam output is more than the process steam requirement, the steam can be fed to an extraction cum condensation turbogenerator. Whatever steam is required for the process, it is withdrawn and the balance steam will be utilized in the power generation.

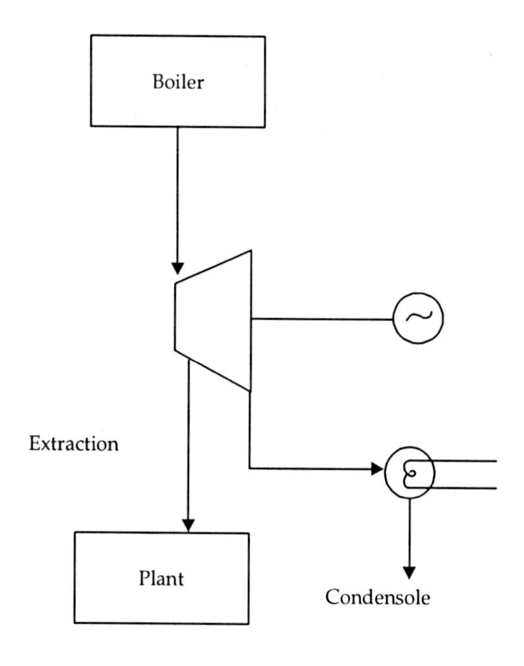

Scheme II

All the bagasse produced will be used up in the boiler and for off season power generation, an alternate fuel like other biomass, lignite or coal is to be procured.

In case of a new plant, cogeneration should be given due attention in the design stage itself and process parameters and boiler design must be finalised with the cogeneration included.

Three key technical aspects must be given due consideration for the improved profitability of a cogeneration scheme in a sugar mill.

MOISTURE CONTENT IN BAGASSE

Bagasse as it produced contains approximately 50% moisture and has a calorific value of 1800 Kcal/kg. If bagasse is used as it is, the amount of heat made available for steam generation will be less. Reduction of moisture content in the bagasse has a twofold advantage

i) As shown in Fig. 1, the net calorific value of bagasse increases by 240 Kcal/kg for every 5 point decrease in the moisture percentage.

ii) The demand for excess air for complete combustion diminishes from about 50–60% to 20%. In effect wastage of heat generated in heating the excess air to the furnace temperature, is reduced.

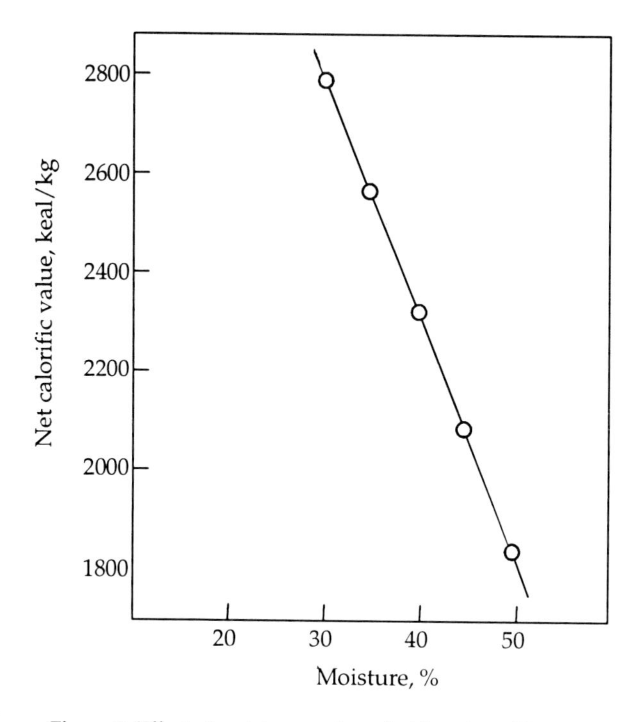

Figure 1. Effect of moisture on the calorific value of Bagasse

Consequently, the furnace temperature realized will be higher. Effect of moisture content on furnace temperature with excess air as the parameter is presented in Fig. 2. It is evident from the figure, that the temperature obtained with 20% excess air and 30% moisture is more than what could be realized with 50% moisture in bagasse at 50% excess air. Higher furnace temperature means better heat transfer and more heat is transferred to the steam improving its energy content. This effect can be seen in Fig. 3, wherein the amount of heat transferred to the steam at various moisture levels is shown. In other words, for a given amount of heat transfer, consumption of bagasse with reduced moisture content will be lower than that of undried bagasse. Thus reduction of moisture in bagasse results in substantial savings in the bagasse consumption. Bagasse thus saved can potentially generate some more power.

USE OF HIGH PRESSURE BOILERS

Traditionally, sugar factories have been employing low pressure boilers to generate steam. This may be alright if sugar production is the only criterion. But steam at higher pressure and higher temperature allows more power

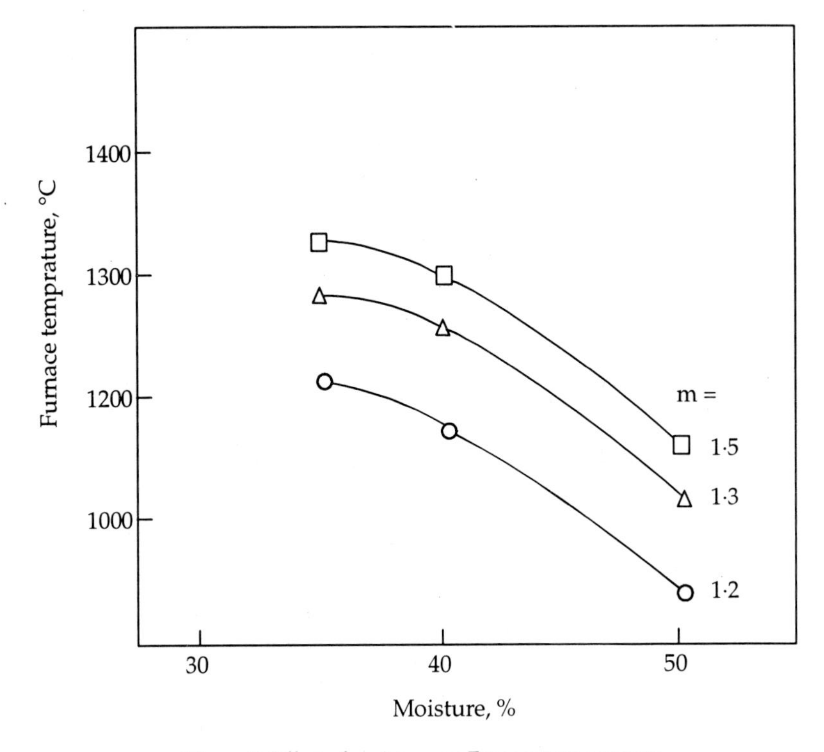

Figure 2. Effect of moisture on Furnace temperature

generation from the same quantity of steam. Table 1 shows information regarding steam inlet condition and turbine power generation, wherein it is evident that higher turbine pressure results in higher power output. For generation of high pressure steam, newer designs of boilers like fluidized bed boilers need to be considered.

Table 1

Steam Condition	kg/sq. cm, °C	Power Output KWH/kg of steam
21	280	91
32	380	118
43	420	135
64	450	152
83	510	174

CONCLUSION

Reviewing briefly the supply-demand position of energy it is suggested

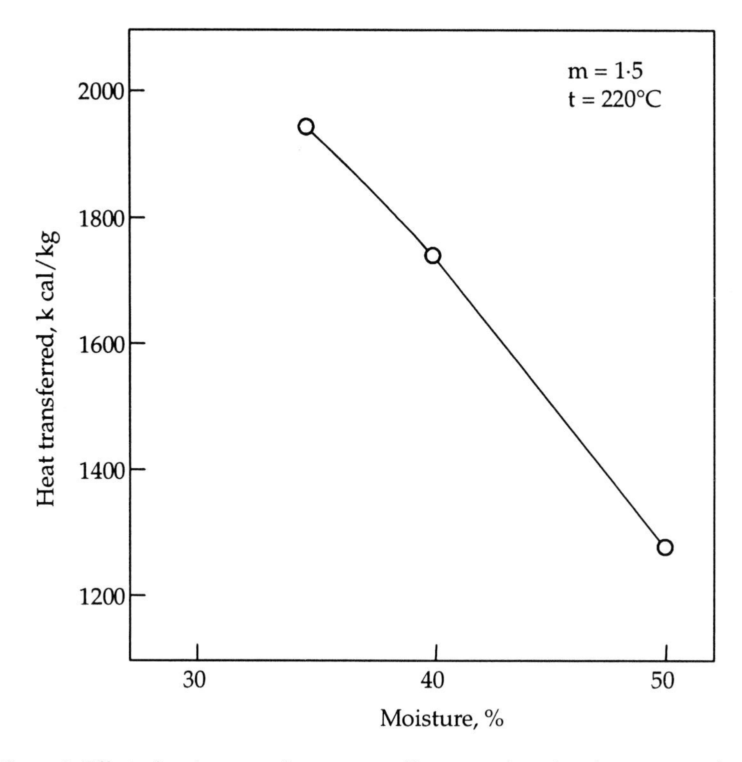

Figure 3. Effect of moisture on the amount of heat transferred to the steam per kg of Burnt Bagasse

that biomass can play a significant role in power generation. Cogeneration in sugar industry is analyzed and three technical aspects—reduction in the moisture content of bagasse, use of high pressure steam and proper selection of operating conditions of the sugar plant, that contribute to higher power generation levels are discussed.

REFERENCES

Boulet, W.P. Waste fuel drying and the energy crisis, *Sug. Jour.*, 8–11.

Furines, J.H. Predrying bagasse using flue gases, *Sug. Jour.*, 39–40.

Maranhao, L.E.C. (1994). Bagasse drying. Paper presented at the ISSCT combined factory/energy workshop.

5

Biomass Energy Plants

Peter C. Rosendahl and Donald Carson

INTRODUCTION

Flo-Sun Energy Corporation is a wholly owned subsidiary of Flo-Sun Sugar Corporation, the largest vertically integrated sugar company in the United States. Flo-Sun farms approximately 180,000 acres of sugarcane and owns three sugar mills in Florida. It has recently completed the construction of state-of-the-art biomass energy plants in partnership with US Generating Company at two of its sugar mills with a combined capacity of 130 MW.

Both energy plants are located within the Everglades agricultural area in South Florida adjacent to the Okeelanta and Osceola sugar mills (Fig. 1). During the sugarcane harvest season (October-April), these plants will burn bagasse—fibrous sugarcane material, and during the off-mill season, wood waste from South Florida urban centres.

The decision to construct what is now considered to be the world's largest biomass plants was based both on enviromental and economic factors (Kohn and Soast, 1994). Burning wood chips and bagasse allows nonrenewable fossil fuels to be conserved and does not contribute negatively to global greenhouse gases. It has previously been suggested that bagasse fuel produces "zero-carbon electricity" which can achieve substantial decarbonization of energy supply (Beeharry and Baguant, 1996). Locating these power plant facilities adjacent to the sugar mills (including a sugar refinery at Okeelanta) assures a dependable supply of energy both as process steam and electricity for maintaining a competitive position well into the future (Fig. 2). Florida Power and Light Company (FPL) purchases this electricity based on FPL least cost alternatives and distributes it within its extensive Florida grid.

The Public Utilities Regulatory Policy Act of 1978 (PURPA) encouraged the independent power industry and co-generation within the United States (Cepero, 1995). This project is an excellent example of the success

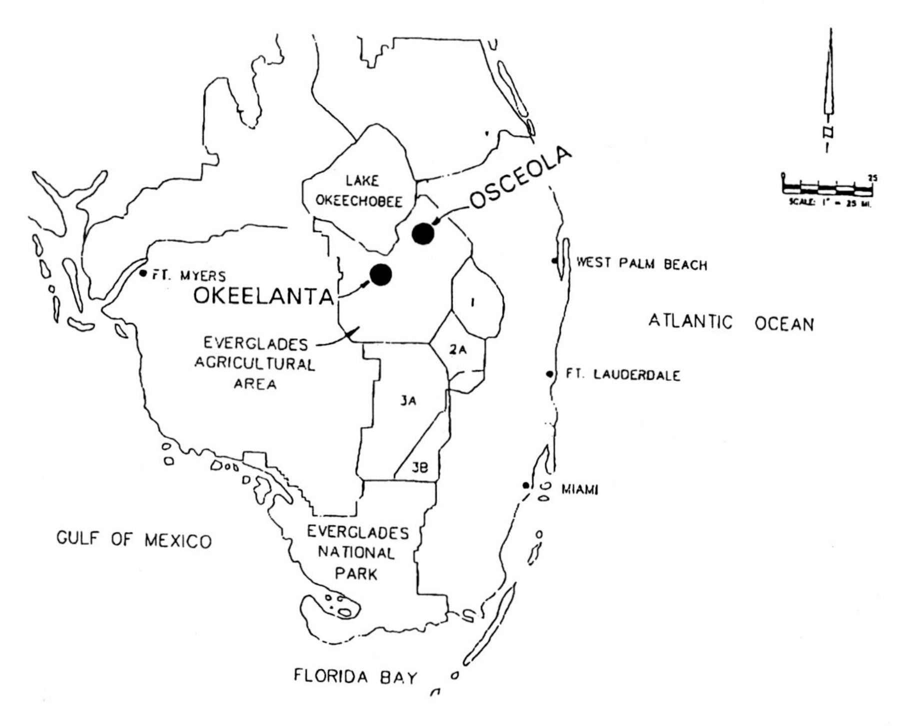

Figure 1.

which PURPA has fostered. While in India, the Ministry of Non-conventional Energy Sources has launched a national programme for promotion of bagasse based co-generation (Government of India, 1994). The Flo-Sun projects described within this paper could serve as a model for future projects in India and elsewhere.

OKEELANTA FACILITY

The Okeelanta biomass plant is located six miles south of South Bay, Florida on US Highway 27 adjacent to the Flo-Sun Okeelanta sugar mill and refinery (Fig. 3). It has a nominal electrical rating of 74.9 MW which is delivered to Florida Power and Light Company via 138 KV transmission lines. The steam plant generates 1,320,000 lb/hr with a pressure of 1,565 psig at 955 degrees fahrenheit. The boiler plant consists of three stoker fired boilers with 440,000 lb/hr each. The turbine/generator is a 3,600 rpm, extraction condensing, down exhaust steam turbine, direct coupled to a two pole synchronous, 13.8 KV, 60 HZ generator. The plant

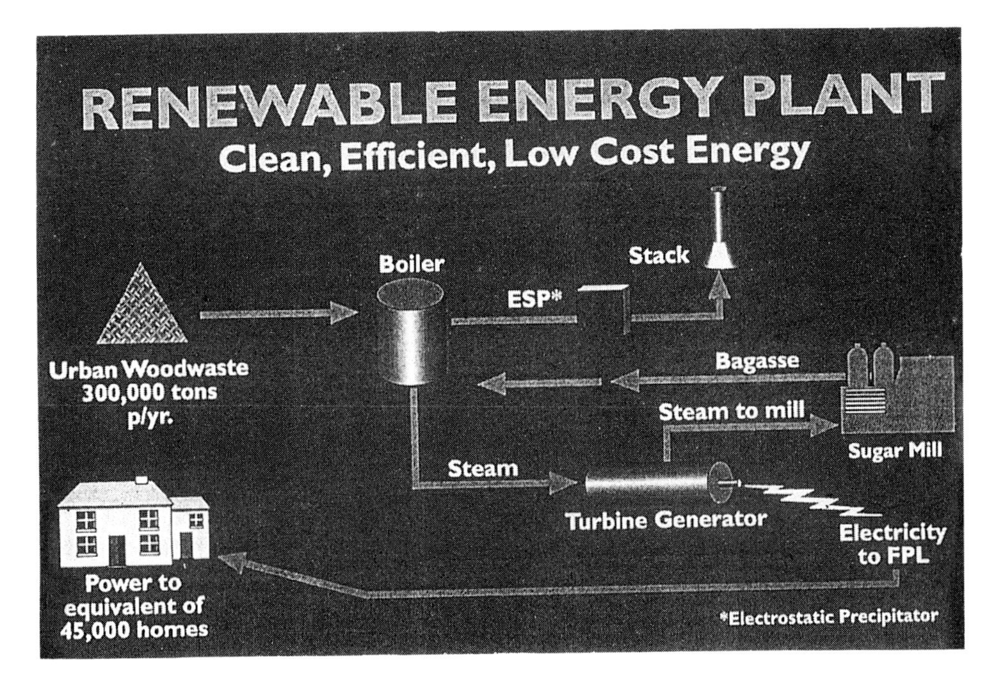

Figure 2.

can export 300,000 lb/hr of process steam, 360 psig at 650 degrees fahrenheit and 420,000 lb/hr, 22 psig at 280 degrees fahrenheit during the sugar mill grinding season.

This facility will handle 5,000 tons/day of bagasse during the grinding season which is delivered via a 270 tons/hr conveyor from the sugar mill. Wood waste is delivered at the rate of 4,200 tons/day from a 150,000 wood yard pile. Wood waste is delivered to the pile via 25 ton trucks, unloaded by two hydraulically operated tractor/trailer dumpers and a self-unloading truck dumper. The heating valves for the dry wood and bagasse are 8,752 and 4,250 btu/lb respectively.

Water is pumped from the Floridian aquifer and treated by pressure filters, reverse osmosis and demineralization. There is a 300,000 gallon filtered water storage tank and a 400,000 gallon demineralized water/condensate storage tank. 90% return condensate from the sugar mill is polished and routed to the demineralized water/condensate tank. A 50% reduction in water usage from conventional sugar mill operation was realized. Waste water from the cooling tower blowdown are routed to a percolation pond which is rotated for crop production annually.

Air emissions are reduced as much as 75% over that of a conventional sugar mill through the use of electrostatic precipitators for particulate removal, selective noncatalytic reduction of NO and mercury reduction

Figure 3.

via activated carbon injection. Plans call for applying the ash as a soil amendment within the adjacent sugarcane fields.

OSCEOLA COMPARED WITH OKEELANTA

The description just outlined for the Okeelanta energy plant also generally applies to the Osceola energy plant with the Osceola plant being smaller. This plant is approximately two-thirds the size of the Okeelanta plant with an output of 55 MW (Table 1). There are two boilers compared with Okeelanta's plant with an output of 55 MW (Table 1). There two boilers compared with Okeelanta's three and due to different site geotechnical conditions, its foundation was built on piles compared with spread footings for Okeelanta. The adjacent mill has a grinding capacity of 13,500 tons/day compared with 20,500 tons/day for Okeelanta.

Table 1. Flo-energy Biomass Co-generation Plants

	Okeelanta	Osceola
Output (MW)	75	55
Boilers (#)	3	2
Foundation (Type)	Spread Footing	Piles
Cooling Tower (Cells)	3	2
Sugarcane (Short Tons)	2,850,000	1,850,000
Mill Grinding Capacity (Tons/Day)	20,500	13,500
Fuel (Tons/Year)	1,025,000	635,000
—Bagasse	705,000	425,000
—Wood	320,000	210,000
Water Consumption (MGD)	2.4	1.5
Ash (TRY)	15,000	10,000
Equivalent Residential Customers	45,000	27,000

Osceola's annual fuel requirements are 635,000 tons consisting of 425,000 tons of bagasse and 210,000 tons of wood while Okeelanta burns, 1,025,000 total tons containing 705,000 and 320,000 tons of bagasse and wood respectively. These fuels should generate approximately 15,000 and 10,000 tons of ash. Water consumption is 2.4 and 1.5 MGD for the Okeelanta and Osceola power plants. Their facilities generate enough electricity to satisfy the needs of approximately 75,000 residential customers.

CONCLUSION

The synergy realized by placing renewable biomass energy plants adjacent to Flo-Sun sugar mills can best be illustrated in Fig. 4. Both a food grade mill and, at other times, a conventional mill and refinery are powered by steam and electricity generated at the adjacent renewable energy power plant. Air emissions are eliminated at the sugar factories and significantly reduced at the new power plants. Urban centres such as Miami are now

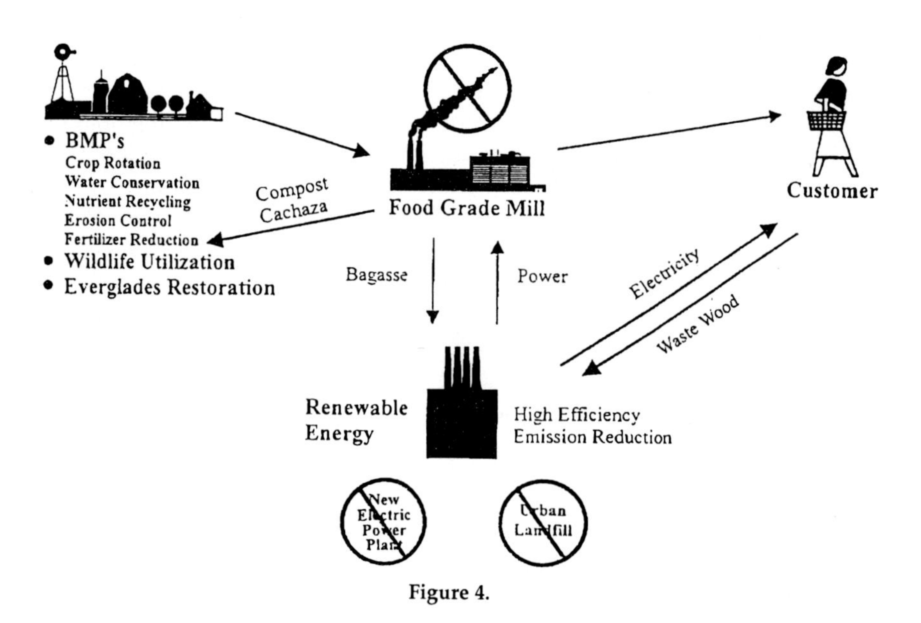

Figure 4.

linked to sugar farming through electricity import and wood waste export. New value added sugar products are supplied to the ever expanding adjacent urban centres.

The sugar factory has become a sustainable component making efficient use of natural resources and achieving a balance between economic vitality and environment. The need for new costly and environmentally impacting landfills in South Florida are diminished, as well as avoiding the immediate need for new power plant sites. Enhanced values are passed along to the customers who also benefit from the continued development of new sugar products made possible by this integrated, profitable sugar/energy/ agricultural complex.

REFERENCES

Beehary, Revin, Panray and Baguant, Jawaharlall. (1996). Electricity from Sugarcane Biomass for Decarbonization of Energy supply in Mauritius. The Environmental Professional, Vol. 18, Special Issue No. 1, July. Olympia, W.A.

Cepero, Gustavo R. (1995). Flo-Sun's Cogeneration Programme Sugar Y. Azucar. July, Englewood Cliffs, N.J.

Government of India. (1994). National Programme On Bagasse Based Co-Generation, Ministry of Non-Conventional Energy Sources. January, New Delhi.

Kohn, David and Soast, Allen. (1994). Plants Will Burn Biomass Cleanly, *Engineering News Record*. December 12, McGraw-Hill, Inc.

6

Thermal Conversion of Biomass

Alex Green, Mauricio Zanardi and Sergio Peres

BIOMASS

Photosynthesis in plants, powered by sunlight, can link glucose molecules into storage compounds (starches) or structural compounds (cellulose). Biomass is made by photoreactions such as

$$6CO_2 + 5H_2O \xrightarrow{\text{Sunlight}} C_6H_{10}O_5 + 6O_2$$

The reaction leads to production of cellulosic type compounds. This equation implies that 264 kg of carbon dioxide and 90 kg of water are consumed when producing 162 kg of biomass while releasing 192 kg of oxygen. Biomass, thus, represents stored solar energy. When biomass is burned for energy it simply releases carbon dioxide and water that it took out of the atmosphere when it was produced. Thus it is effectively CO_2 neutral.

Combustion of biomass has been use by humankind for 500 millennia to generate heat and light. Biomass was the world's predominant energy source until fossil fuels took over in the industrial world, during the industrial revolution. Unfortunately, fossil fuels when burned release carbon dioxide extracted from the atmosphere in an earlier geological era. Biomass as a renewable energy source, can now re-assume greater energy loads while serving a number of other economic, environmental and social purposes. Geographic areas with abundant sunshine and rainfall are particularly suited for increased biomass energy use.

Here biomass would include wood, wood waste, agricultural waste, energy crops, municipal solid waste, sewage sludge and cellulosic type industrial waste. Biomass is widely available as residue from wood and food processing industries in the form of sugarcane bagasse, sawdust, slags, chips or as municipal solid waste (MSW). The pulp and paper industry has extensive experience in biomass use for process heat while disposing of waste forest products.

THE UNIVERSITY OF FLORIDA AND BIOMASS

Florida with its abundant sunshine and rainfall is probably the most favorably positioned region of mainland USA for using the energy potential of biomass. The University of Florida (UF), as its land grant school, has a longstanding programe on biomass energy. Among the favourable energy crops investigated are the tropical leguminous shrub, leucaena and perennial tropical tall grasses such as elephant grass, sugarcane, and energycane that are well adapted to the humid areas with long growing seasons and high rainfall (Prine and Woodword, 1994). Eucalyptus varieties are typically the most productive woody species on lands available for energy cropping (Rockwood et. al., 1993). These studies are a part of the University of Florida's broad and longstanding effort to generate combustible gas from biomass (Smith and Frank, 1988).

THE UNIVERSITY OF FLORIDA–SRI VENKATESWARA UNIVERSITY COLLABORATION

Thermal applications of biomass can be traced to times before recorded history, and it would be impractical, in a finite paper, to provide references even to major studies. Instead, we will summarize and adapt the work of the UF Clean Combustion Technology Laboratory (CCTL) and the UF collaboration with Sri Venkateswara University (SVU) directed towards advancing the use of bioenergy in India and the USA. These CCTL publications contain extensive references to the related literature. In contrast to the mainstream activity of designing facilities to fit available fuels the CCTL has concentrated on blending hydrocarbon fuels to fit existing facilities. These began in connection with an interdisciplinary study of *Coal Burning Issues* (Green, 1980). The objective then was to assist Florida utilities in reducing their use of oil by increased use of coal. What evolved was a general focus on fuel blending for various energy efficiency, environmental and economic objectives. Fuel blending in boilers is now receiving considerable attention in the USA and European Union (EU) especially fuel blending in gasifiers. For a variety of reason it might be expected that the technical literature on fuel blending will soon grow rapidly since it broadens the available environmentally advantageous solutions to problems related to conversion of hydrocarbons to energy.

CO-COMBUSTION AND CO-GASIFICATION

Co-combustion, the conferring of two or more fuel types in a single combustor, can be helpful in achieving easy start-up flame stability over a large range of inputs, greater combustion efficiencies, reduced emissions and other technical and environmental advantages. In 1980, a research and development program was initiated leading to the book *An Alternative to Oil, Burning Coal with Gas* (Green *et al.*, 1981). The work pointed to a combination of domestic fuels that could be used in available boilers designed for oil when oil prices were high. As the work progressed, the CCTL considered more general technical aspects of confiring of all domestic fuels, including biomass, for energy, environmental and economic objectives. In 1988, this work received a United States Departmental of Energy National Energy Innovation Award and a Florida Governor Energy Award.

With the emergence of gas turbines (GTs) as the most efficient of transforming heat to mechanical energy, the CCTL has also directed its attention to advancing the science and technology of fuel blending in gasifiers with the objective of producing a gaseous fuel that can be used with advanced gas turbines (Green, 1986). Environmental objectives included utilizing biomass, that is CO_2 neutral to help mitigate greenhouse gas emissions (Green, 1989). To a considerable extent, this thrust is a form of defense conversion, since combustion or gas turbine has its origin as an aircraft engine during World War II, and was greatly advanced in response to aircraft performance requirements driven by the Cold War. The application of GTs to peacetime service for power generation is one of the most promising examples of "swords to plowshares" that may evolve from the end of the Cold War (Green and Chernyy, 1995).

Unfortunately some problems arise in using biomass as sole gasifier feedstock: (1) the capital and operation and maintenance costs of gasifiers are high, especially those derivative of coal programs, (2) cultivated biomass feedstock in most locations costs more than coal, which has been used in gasifiers for almost two centuries, (3) biomass availability is seasonal, (4) low biomass densities limit the economic transportation range of biomass, (5) high oxygen to carbon ratios of biomass lead to gaseous products with low heating values, (6) green biomass requires drying, (7) eutectic properties of biomass ash can lead to low melting points, (8) the science of biomass gasification is still too primitive for predicting results of various proposed gasifier designs, and (9) testing first of its kind, full scale gasifiers, is very costly.

The CCTL R&D program addresses these problems by (1) blending biomass with other domestic fuels, (2) focussing on indirectly heated gasifiers, (3) seeking additional societal functions that can be served by gasifiers, (4) seeking valuable storable by-products that can be produced by gasifiers, (5) bringing modelling techniques from other disciplines to help develop a predictive science of gasification, and (6) conducting laboratory scale studies to verify the proposed predictive models. The remaining sections of this article present a strategic analysis of near term thermal conversion technologies using biomass, usually blending with other fuels. The analysis considers: (1) Fuels and fuel blending, (2) Combustors, (3) Pyrolysis, (4) The science of pyrolysis, (5) Gasification, (6) Cogeneration, (7) Liquids and Chemical Production, and (8) Conclusions on Optimum use of Biomass.

Fuels and Fuel Blending

The properties of fuels important in confiring or co-gasifying are largely dictated by the pyrolysis, physical, chemical and other basic characteristics of the available fuels and the available or planned combustion system. The hydrogen/carbon atomic ratio (H/C) and the oxygen/carbon (O/C) ratio are important variables since they strongly influence the physical phase of the feedstock or the products of solid fuel pyrolysis. Thus for non-oxygenated hydrocarbons (O/C = 0) low H/C ratios (H/C≤1), usually corresponds to solids and liquids, and high H/C ratios (H/C≥2) to liquids and gases. However, molecular bonding plays a role and the gas, acetylene (a), and the volatile liquid benzene (b) have H/C = 1. Figure 1A shows y = H/C and x = O/C atomic ratio coordinates of potential solid fuel feedstocks. The biomass points are high, medium and low volatile averages of wood, agricultural residues (Agric), bark, and energy crops (E crops) drawn from the extensive thermal data collection of Gaur and Reed (1995). The coal points are data for lignite (LIG), sub-bituminous (SUBB), high volatile bituminous coals (HVB), middle volatile bituminous coal (MVB), subanthracite (SUBA), and anthracite (ANT) drawn from *Steam* (Babcock

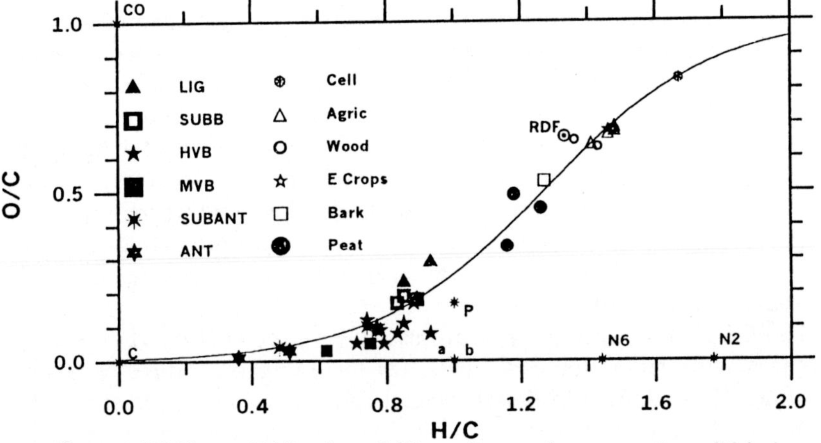

Figure 1. (A) The y = H/C and x = O/C coordinates of representative solid fuels.

and Wilcox Inc., 1982), *Combustion* (Combustion Engineering, 1980) and other sources. These hydrocarbons lie near a smooth "coalification" curve that can be represented by the logistic function.

$$x(y)=1/\{1+\exp[(1.27-y)/0.25\}], \; y(x)=1.27 + 0.25 \ \text{in}\ [x/(1-x)] \sim 0.7 + 1.2\,x \qquad (1)$$

The linear expression is restricted to the region between HVB and Agric. In large measure international synthetics fuel programs have been directed towards hydrogenating coals of premium liquid fuels such as the N2 and N6 oils indicated on Figure 1A or fuels with higher H/Cs (Probestein *et al.*, 1982). Oxygenation can also convert solid hydrocarbons to liquids of gases (e.g. C + O → CO). However, heating values are lowered by oxygenation.

The volatile content of hydrocarbon fuels is a major parameter determining ease of gasification of a solid fuel. Figure 1B shows the weight

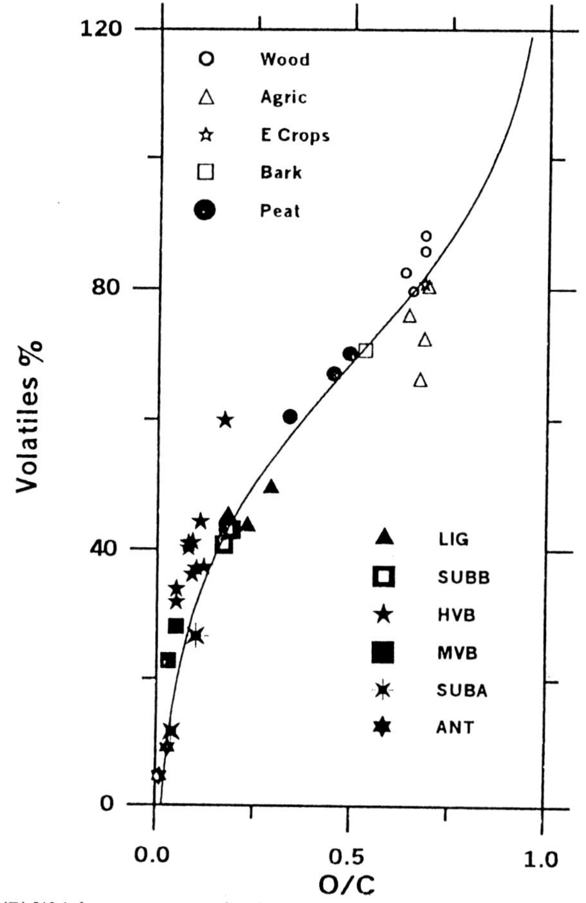

Figure 1. (B) Weight percentage of volatiles vs. x = O/C ratio of representative solid fuels.

per cent of volatiles from representative solid fuels vs. x = O/C. The high volatile content of biomass affords a great advantage in gasification. Bark, lignin and peat are next favored followed by lignite, subbituminous and high volatile bituminous coal. The volatile yields from most bituminous coals are lower and the volatiles from anthracite are only of the order of 10% or less. The smooth logistic-like curve uses Eqn. 1 together with an excellent representation of $V(y)$.

$$v=100\,[1-\exp(-(y/d)]/[1+\exp(y_0-y)/d] \text{ where } y_0-0.92 \text{ and } d = 0.32 \quad (2)$$

High volatility generally is associated with low heating values of the feedstock and the pyrolysis gases. Figure 1C gives the higher heating values (left scales in kJ/g, right scales in MBTU/1b) vs. O/C. While biomass has high volatility the heating values of the pyrolysis gases are marginal. On the other hand, coal has poor volatility but good heating values. The increasing O/C values (see Figure 1A) explain why heating values declines as O/C increases. This behavior can be derived from Dulong's formula that may be written as

$$HHV=33.8\,C_f + 144.3\,H_f-18.0\,O_f + 10.0\,S_f-1.5\,N_f-2.1\,Ash_f \text{ in kJ/g} \quad (3)$$

where C_r, H_r, O_r, N_r and Ash_r are the mass fractions of carbon, hydrogen, oxygen, sulfur, nitrogen and ash respectively in the fuel. For pure CH_yO_x solids, we have found by least squares an improved quadratic relation that can be expressed as

$$HHV\text{-}32\{1+0.369\,z-0.137\,z^2\}/\{1+(y/12)+(16x/12)\} \quad \text{where } z=y-0.85x \quad (4)$$

The solid curve in Figure 1C is obtained by using the $y(x)$ function of Eqn. 1 in Eqn. 4. The scatter is probably due to the varying sulfur and ash content in our coal biomass data base.

The conversion of heat to mechanical work can be accomplished via hot air engines (West, 1986), steam engines or internal combustion engines (ICE). Hot air and steam can be made by the direct combustion of biomass. Reciprocating forms of ICEs began taking over from steam engines at the the beginning of the 20th century. In the last two decades, however, combustion turbines (CT) or gas turbines (GT) have emerged as the most efficient converters of fuel to mechanical energy. For ICEs or combustion turbines to be powered by biomass or any solid fuel, the biomass must first be converted to liquid or gaseous fuel.

Combustors

Biomass combustion can result in the transfer of 65–80% of the heat content of the wood into hot air, steam or hot water. In direct combustion, the fuel is placed directly in the furnace or boiler. Most of the biomass used for

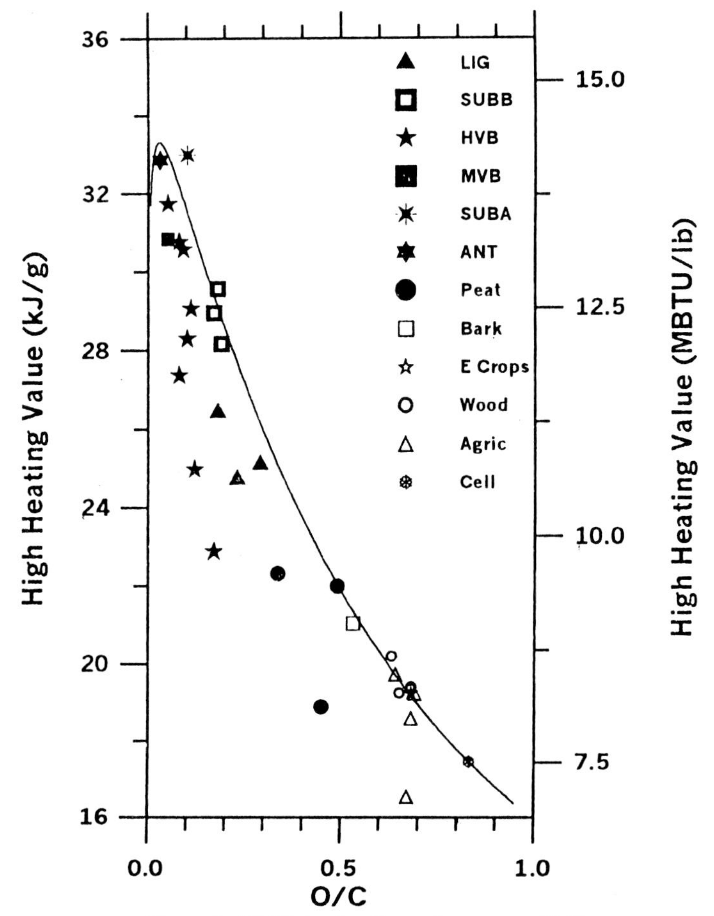

Figure 1. (C) Higher heating values vs. x = O/C

energy today is combusted directly in grate fired systems, suspension burners or fluidized bed systems. In most pulp and paper and food processing industries the heat released is converted for use as process steam.

Great Fired Combustors

These are currently the most widely used combustors and include both pile burners and spreader-stokers. Pile burners can be of the following types: (a) Dutch ovens; (b) Wellons cells; (c) Lamb-cargate wet cell; and (d) Inclined grate systems. Spreader-stokers are distinguished by their feed mechanism that may use horizontal or sloping grates; stationary or moving, air cooled or water cooled. Grate fired units may be refractory lined or waterwall furnaces and can handle fuels with high moisture contents (MC). Refractory lined systems are capable of handling fuels with moisture content greater than 60% in large chips and can stage the combustion very effectively.

Suspension Burners
Suspension burners require biomass fuels with very low moisture content (less than 15% MC) and fuel particles sizes less than 2 mm. Thus, the fuel preparation and storage cost are greater than for grate fired systems. Suspension burners may be of interest if finely divided fuel is naturally available, for example sawdust. Otherwise the high grinding costs and low fuel handling capacity limit the usefulness of suspension burners.

Fluidized Bed Combustors (FBC)
Fluidized bed combustors, which have the greatest versatility and potential for emission control can use either bubbling or circulating beds. Both use air to fluidize the bed particles, which can be an inert material such as sand, a sulfur sorbent material (limestone or dolomite) or a sand/sorbent mixture. The combustor chamber contains a high mass of inert material capable of absorbing energy from fuels that combust in a highly turbulent fashion. The bed media contains sufficient energy to successfully burn fuels with over 60% moisture content.

Pyrolysis

Pyrolysis, the decomposition of organic matter at high temperature in an inert atmosphere or vacuum, has been applied for hundreds of years in the production of charcoal that has many domestic, metallurgical and chemical uses. The main products of biomass pyrolysis are: charcoal, pyroligenous liquid and gaseous products. The amount of each depends on the chemical composition of the biomass used and operational conditions. In charcoal, production typically, 100 kg of dry wood will produce 30 to 50 kg of charcoal, 35 to 50 kg of pyroligneous oil and 15 to 20 kg of gases. However, the yields and chemical compositions of the liquid, gas and charcoal change with the pyrolysis temperature, residence time, pressure, feedstock and other variables.

Pyroligneous Liquids
Pyroligneous liquids typically consist of 80% pyroligneous acids (wood spirit) soluble in water and 20% of wood tar containing phenolics. Pyroligneous acids (wood spirit) soluble in water and 20% of wood tar containing phenolics. Pyroligneous liquids are considered to be feasible substitutes for industrial fuel oil. Their high heat value and their acidity make them corrosive to mild steel but much less to stainless steel, copper and ceramics. The most valuable commercial products which can be obtained by processing pyroligneous liquids are: heating oil, creosotes, gualacol and froth flotation oil, wood preservative, and insulation products.

Wood Gas

The main combustible components of wood gas are: CO, CH_4, H_2 and C_2H_4. The main non-combustible gases are CO_2, H_2O and N_2. The relative concentrations of these components vary according to the types of reactor used and the terminal temperature. In reactors where heat is provided by combustion of part of the wood in air, the CO_2, H_2O and N_2 concentrations are substantial. Because of low heating values, wood gases are usually fired directly on site to dry and heat the material in a carbonization plant, or to generate steam. When used as a fuel for engines, the gas has to be cleaned to remove the liquid droplets and particulates.

Charcoal

Important properties of charcoal are: (a) yield: ratio of the charcoal weight to the dry input material; (b) content of volatiles: the weight loss per unit weight of charcoal when heated at 900°C (1600°F) under vacuum or in an inert atmosphere; (c) fixed carbon content: the dry charcoal weight minus the weight of volatiles and incombustibles (ashes) per unit weight of charcoal.

Pyrolysis Reactors

The type of reactor determines the pyrolysis processes and dictates the experimental conditions, the particle size, and also the yield and composition of the different pyrolysis products. The main traditional types of pyrolysis reactors are:

(A) Kiln for producing only charcoal;
(B) Converts for pyrolyzing small particle size feedstocks;
(C) Retorts for pyrolyzing billets or logs up to 1 foot in length and 7 inches in diameter;
(D) Rotary kilns for generating, char, oil and gas from various feedstocks.

The Science of Pyrolysis

Pyrolysis is the initial step in combustion, gasification or conversion to liquid of any solid fuel. From a predictive standpoint, it is unfortunate that the science underlying pyrolysis is still in a state of great uncertainty. For example, reaction rates reported in the literature for mass loss of a specified solid particle at a specified temperature often differ by two or three orders of magnitude. Members of the CCTL have devoted considerable efforts towards developing a predictive science of pyrolysis. The need is great since the amount and nature of the pyrolysis volatiles is a key issue in biomass gasifiers although to a lesser extent in coal gasifiers. Pyrolysis volatiles in the form of carbon dioxide and water vapour cannot serve directly as energetic fuels. On the other hand, molecular hydrogen, methane and to a lesser extent carbon monoxide are good GT fuels. Unfortunately, the literature on the composition of pyrolysis gases is sparse, fragmentary, inconsistent and very difficult to systematize. In an effort to arrive at the general systematics of gaseous emissions we have resorted to molecular modelling of pyrolysis processes such as was used in our early CCTL co-firing studies (Green and Pamidirmukkala, 1982, 1984, Green *et al.*, 1984; 1986, 1989). We assumed that as temperature increases coal depolymerizes through various intermediate states somewhat like the decay of radioactive chains (Green, 1955). With this methodology we were able to accurately incorporate mass loss data into our kinetic models of coal, coal water mixtures (CWM) and CWM-natural gas combustion. In our current cogasification studies, we require a more detailed pyrolysis approach that gives specific gaseous yields. Here, to facilitate the use of the sparse experimental data, we employ the integral logistic function (ILF) for yields and the differential logistic function (DLF) for rates

$$Y = W \exp X/(1+\exp X) \text{ and } R = dy/dT = (W/d) \exp X/(1+\exp X)^2 \quad (5)$$

where $X = (T-T)/d$. We have also found it helpful to recast the Arrhenius rate equation in terms of the dimensionless quantities $u=(1000/T)$ and $q=E/1000$ R using.

$$k = A\,T^{n} \exp{-E/RT} = Fu^{n} \exp{-q}\,(u-1), \text{ where } F = A\,(1000)^{n}\, exp{-q} \quad (6)$$

In this $[F, q, n]$ parametrization the magnitude parameter F is the reaction rate in sec^{-1} at 1000 K, a somewhat central temperature for combustion and gasification phenomena.

In applying a molecular decay model, we first make a transformation from the independent time variable to temperature that is applicable for constant heating rates ($T=at+b$) as used in many thermal measurements. In such cases, derivatives with respect to time in first order processes become derivatives with respect to temperature simply by letting $Dt=a\,dT$. Figure 2A shows a decay scheme for the pyrolysis of sweet gum borrowed from nuclear decay theory. The SG denotes the initial sweet gum hardwood molecule that under increased temperature decays at a rate k_1 to a gas G_1, a liquid L_1 a tar T_1 and a char-ash X_1. The liquid (L_1) pyrolyzes further with a rate constant λ_1 to a gas G_2. The tar (T_1) pyrolyzes at a rate μ_1 into a tertiary gas G_3 and a final T_2. The char X_1 also can undergo further pyrolysis at a rate *ov* v_1 into a gas G_4, a final char-ash X_2. In most analyses of coal data additional decay steps are needed to fit available data.

In the CCTL methodology all experimental gaseous emission rates vs. temperature data are fit in terms of a minimal set of differential logistic functions or all yield vs. temperature data in terms of a minimal set of integral logistic functions. Letting D denote d/dT gaseous emission rate equations for all decays leading to gases are represented in terms of the same logistic components. Thus, we assume that each of the gases G_1, G_2,

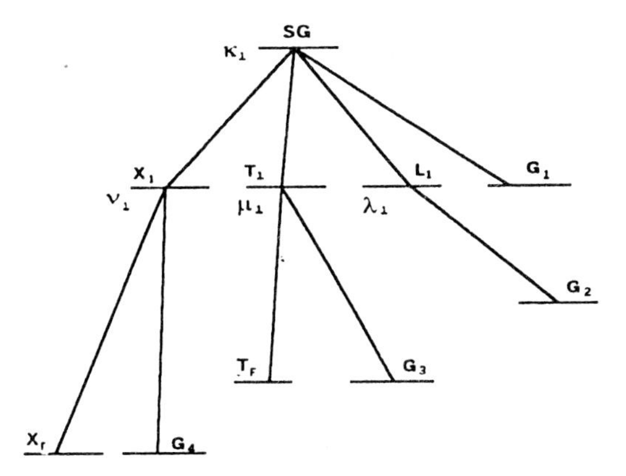

Figure 2. Molecular model results for sweet gum hardwood **(A)** Molecular decay scheme.

initially represents a single composite gaseous molecule that instantly breaks up into the observed small molecules CO_1, CO_2, CH_4 and H_2 having the same values of *Ti* and *di* but with molar proportion determined by their Wi's. The W values for the *Y*'s and *R*'s composite for the *G*s are all taken as unity in Eqn. 5. Then, the rates of decay for all state numbers ($S=B$, biomass, *L*, liquid; *T*, tar or *x* char ash) take the form.

$$DS = R_I - R_J \tag{7}$$

When the differences $R_I - R_J$ are integrated the constants of integration can be set to zero and the $R_I - R_J$ becomes $Y_I - Y_J$. After giving attention to initial conditions, we can, thus, very simply integrate all differential equations by using this molecular decay model. Then, by algebraic manipulation we can solve for the reaction rates with all results coming out in the form

$$\mu_i = R_J/(Y_I - Y_J), \text{ with } Y_0 = 1 \tag{8}$$

Having found a molecular model, identified a decay scheme, solved the set of differential equations, we can now generate all the reaction rates as functions of temperature using Eqn. 8. We next transform back from temperature to time by multiplying the rate constants using a realistic heating rate such as 1000 K/sec rather than the measured rate in the thermal experiment that is usually in the 5–100 K/min range. Then, with our and the reaction scheme illustrated in Fig. 2A and the temperature-time profile kinetic code $T = at + b_1$ we can calculate the gaseous emission rates via the usual kinetic rate equations. When we approximate the rate constants derived from Eqn. 8 by simple Arrhenius forms using

$$F = (1/2d) \exp (T_J/2d) (1 - T_J/1000) \text{ and } q = T_J^2 (2000d)$$

We usually generate emission rates close to the input rates. If we use the non-Arrhenius reaction rates (Eqn. 8) directly, we reproduce the logistic fits to the experimental rates or yields.

To illustrate this approach, we have fit the experimental yield data on sweet gum-hardwood obtained at MIT, by Nunn *et al.* (1985), with minimal sets of logistic functions, and have derived the emission rates and the reaction rates using our molecular model. Figure 2B assembles the results of our fits of the integral logistic sums to the yields of CH_4, C_2H_4 and C_2H_6. Figure 2C shows the yield data and logistic fits to the CO, CO_2, CH_2O and H_2O yield data. A base set of only four logistic functions are required. The derived rates of release of gases from sweet gum hardwood are given in Fig. 2D. The derived reaction rates for the decay steps are shown in Fig. 2E. The logistic parameters are given in the Table in Fig. 2F. It should be noted that the volatiles produced in the Nunn *et al.* (1985) study only total to 47%.

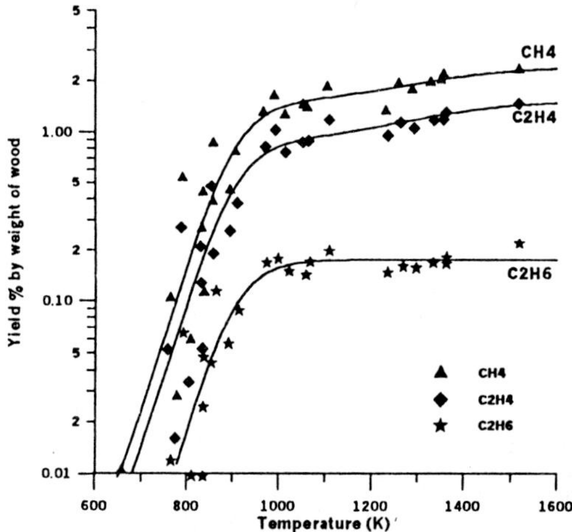

Figure 2. (B) Gaseous yields for CH_4, C_2H_4 and C_2H_6.

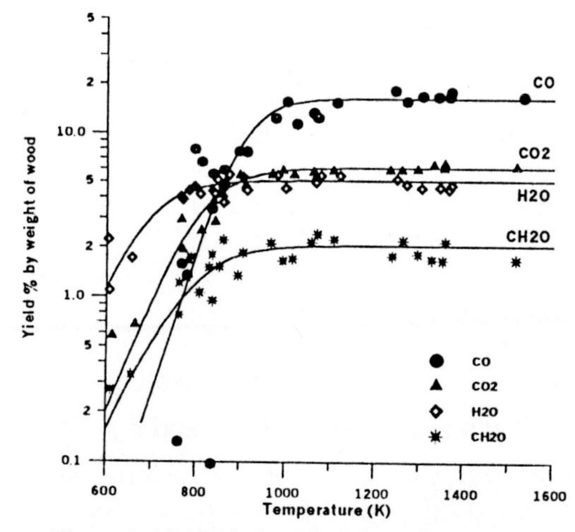

Figure 2. (C) Yields for CO, CO_2, H_2O and CH_2O.

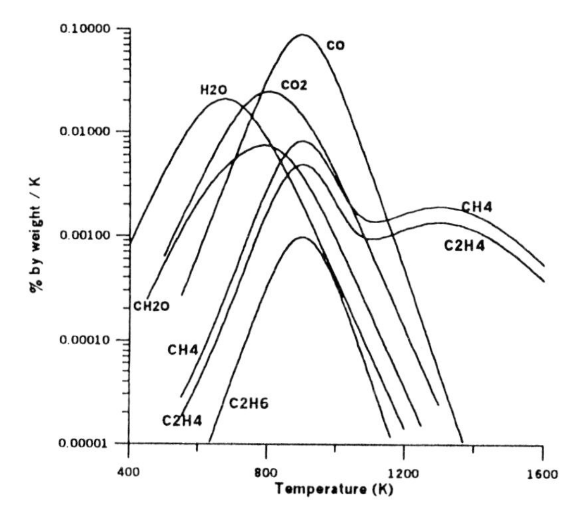

Figure 2. (D) Rates derived from B and C.

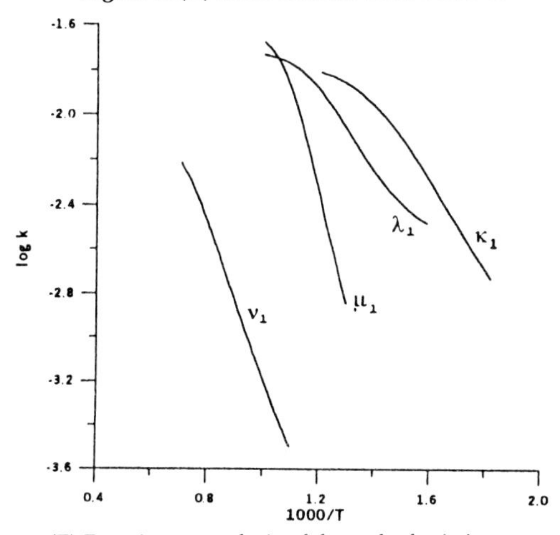

Figure 2. (E) Reaction rates derived from the logistic representation.

species	w_1	T_1	d_1	w_2	T_2	d_2
H_2O	5.03	675	60	0.05	800	60
CO_2	5.77	800	60	0.33	900	45
CO	0.47	800	60	15.9	900	45
CH_4	1.44	900	45	0.91	1300	120
C_2H_4	0.85	900	45	0.65	1300	120
C_2H_6	0.18	900	45	-	-	-
CH_2O	0.45	675	60	1.61	800	60

Figure 2. (F) Table of the logistic parameters.

According to Borenson *et al.* (1989), when the tar is also cracked, the total volatiles add to 76%.

Biomass Gasification

Gasification is a thermal process of changing a solid fuel such as coal, biomass or municipal solid waste into combustible gas and oil vapours. Pyrolysis is a form of gasification, however, gasification usually involves some secondary high temperature chemistry that improves the heating value of gaseous output and increases the gaseous yield. Four basic internal combustion-gasifier types have evolved for biomass over many years. These are illustrated in Fig. 3 and described below. Generally, these produce a low heating value gas because of the nitrogen from the input air and the CO_2 and H_2O combustion products. When oxygen is used instead of air for combustion, dilution with atmospheric nitrogen is minimized. Indirectly heated gasifiers separate the combustion to generate the heat for gasification from the chemical reactor function that transforms the solid fuel into a combustible gas. This minimizes dilution of the product gas with nitrogen and with the carbon dioxide and water vapour products of combustion. Brief descriptions of IHGs drawn from the literature follow those of directly heated gasifiers.

Fixed-Bed Downdraft Gasifiers

In downdraft gasifiers, as shown in Fig. 3A, the fuel is fed through the top of the vessel, and combustion air is fed in the "throat" below the combustion zone. The product gases are drawn through this "throat" and out through the annular shell. As the product gases pass the combustion zone to the reduction zone most of the tars and oils in the gas are cracked to more manageable light hydrocarbons, and the gases typically exit at 700°F (371°C), thus maintaining light hydrocarbons in the vapour phase. Because of the complexity of the annular shell and throat assemblies, current downdraft designs have not been scaled up beyond 12 MMB tu/hr (3.5 MW). The cleaner gases from downdraft systems are easily close-coupled with internal combustion engines.

Fixed-Bed Updraft Gasifier

Figure 3B is a typical design with a refractory-line carbon steel cylinder having an air grate at the bottom of the vessel, and a gas outlet, and a gas-tight fuel feeding port at the top of the vessel. Air is fed upward through the grate into a combustion zone to generate the heat for the gasification reactions. The rising gases next encounter a reduction zone where char reacts with the carbon dioxide formed in the combustion zone to form carbon monoxide. Next in the devolatization zone the hot gases rising through the bed drive off hydrogen, light hydrocarbons, and some tars. The hot product gases then pass through the incoming fuel drying zone

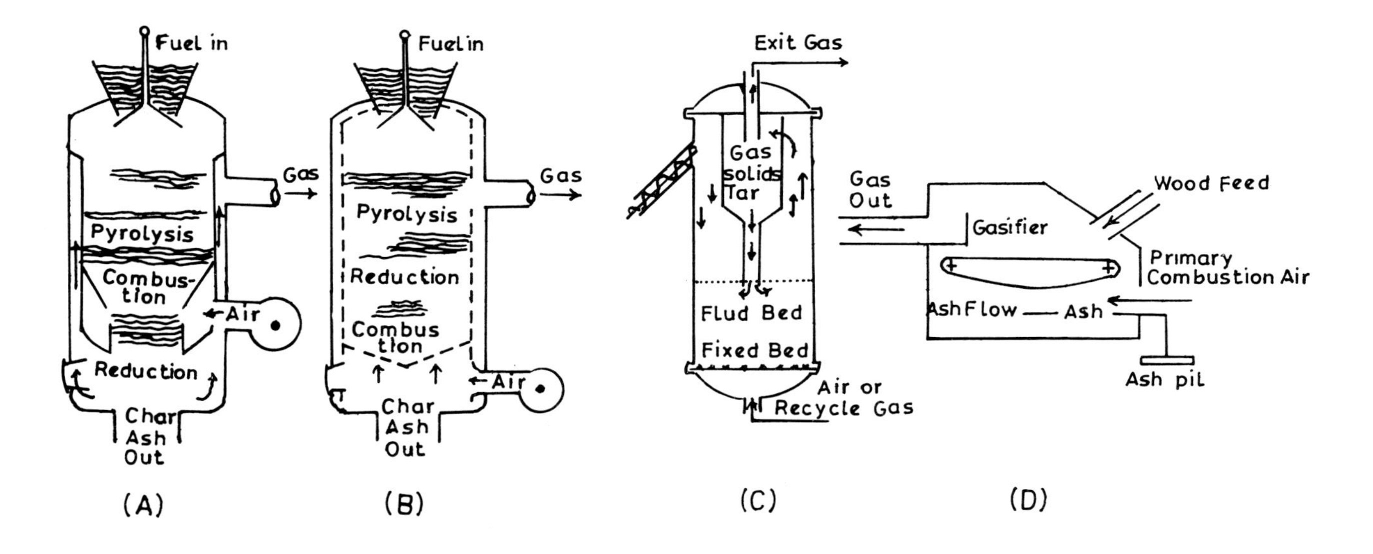

Figure 3. Combustion—Gasifiers. (A) Fixed-bed downdraft gasifier. (B) Fixed-bed updraft gasifier. (C) Fluidized bed gasifier. (D) Moving bed gasifier. (adapted from Reed, 1981 and Florida Power Corp., 1985)

before exiting the vessel at temperature from 170 to 250°F (76.7 to 121°C). Tars and oils formed at high temperatures in the reduction and devolatization zones tend to condense at the lower temperatures of the gasifier exit. Either the gas has to be heated to prevent condensation or the liquids must be recovered as fuel. Experience with updraft systems indicates a maximum output of roughly 100 MMB tu/hr (29.3 MW) of low heat value gas.

Fluidized Bed Gasifiers

These designs (Fig. 3C) are similar to fluidized bed combustion systems. They have inherent feedstock flexibility because of the fluid nature and temperature retention of the bed. Other advantages include ease of scale-up, higher gas exit temperatures, feedstock flexibility, and good turndown ratio (typically on the order of 3:1). Current commercial designs for large fluidized bed systems are available with outputs greater than 200 MM B tu/hr (58.6 MW). Problems develop with particulate loading in the exit gas stream, start-up bed heating and parasitic loads.

Moving Bed Gasifiers

These gasifier designs (Fig. 3D) are very similar to travelling grate combustors and incorporate proven materials handling systems that carry fuel through the vessel. Reduced oxygen is provided to promote gasification rather than combustion. Design can readily be scaled up but wood gas often causes grate problems.

Indirectly Heated Gasifiers (IHG)

The open literature does not describe details of different types of indirectly heated gasifiers. Such biomass gasifiers may be under development but with information treated as proprietory. Some brief descriptions have been garnered from gasifiers used to produce chemicals (such as methanol) that might be adapted as gasifiers for GT applications.

The Wright Malta Steam Gasifier (Reed, 1981) uses high temperature high pressure steam to convey energy to the gasifier. At the same time the steam serves to reform fixed CO_2 via $H_2O + C- > CO_2 \rightarrow H_2$ yielding finally a medium BTU gas consisting mainly of H_2 and CO_2.

The Battelle gasifier (Sterzinger, 1995) uses two hot chambers separated by a cyclone and circulating sand and char to convey heat from the combustion chamber to the gasification chamber. The hot gas clean up system removes particulates from the output gas before the gas turbine.

Rotary kilns can be used in char and oil generation (Ledford, 1992). By using high temperature, materials and construction for operation at much higher temperature such a system could possibly be used as a biomass gasifier to feed GTs.

The MTCI/Thermochem gasifier (Mansour *et al.*, 1995) provides heat indirectly by using natural gas fired pulsed heaters immersed in the bed of a reactor fluidized with superheated steam.

The Brightstar Synfuels Technology (Smith, 1996) combines steam with biomass at high temperatures and low pressure in an externally heated tubular reformer (similar in many respects to a conventional steam-methane reformer).

The CCTL laboratory scale IHGs (Green *et al.*, 1995, 1996 a,b,c) use electrically heated ceramic, stainless steel or quartz tube systems designed to test the comparative benefits of solid fuel blending, NG-SF blending, steam-SF blending and catalytic and high temperature promotion of gasification processes. Bagasse, leucaena and eucalyptus are among the biomass feedstocks receiving special attention. From an academic research base, "publish or perish" rather than industrial secrecy prevails.

Gasifier Products

In airblown gasifiers the nitrogen component of the air lowers the heating value of the fuel and also potentially enhances NOx formation. Oxygen blown gasifiers avoid nitrogen problems but have high capital cost, due to the oxygen generator, and high CO_2 outputs. Depending upon the gasification agent and the gasifier, biomass gasifiers produce combustible CO, H_2 and CH_4 diluted with non-combustibles N_2, CO_2 and H_2O vapour. The composition of the product gas is determined by the biomass feedstock and gasification agent used, as well as by the operational conditions, such as pressure, temperature, residence time and heat loss or external heat input. In indirectly heated gasifiers, combustion products do not mix with the gasification products and hence CO_2, N_2 and H_2O concentrations are lower. The types of gases produced by biomass gasification can be divided into three categories according to their heat value.

(A) Low Heat Value Gas (~150 BTU/ft^3; 5.6 MJ/M^3): Low energy content gas (LEC) is produced when air is used as the gasifier agent. The gas is used on site since storage and/or transportation of LEC gas are not feasible.

(B) Medium Heat Value Gas (~400 BTU/ft^3; 14.9 MJ/m^3): Medium energy content gas (MEC) is produced when oxygen or steam is used as the gasifier agent or with indirectly heated gasifiers. MEC gas can be used as ICE and gas turbine fuels and for the production of synthetic fuels, such as H_2 gasoline, methanol, synthetic natural gas, etc.

(C) High Heat Value Gas (HHV~750 BTU/ft^3; 27.9 MJ/m^3): High energy content gases are usually produced from MEC gases. These gases can be used as substitutes for natural gas which usually has a heating value of about 1000 BTU/ft^3 (37.2 MJ/m^3).

Co-generation

Several types of gas turbines (GT) for alternative cycles are under development that can be matched with different gasifiers, depending upon whether a low or medium heat value gas is produced.

After hot gas clean-up the synthetic gases can be injected in the combustion chamber of the gas turbine to provide the power to drive the turbine. Approximately two-thirds of the power produced by the turbine is used to drive the front end compressor, that is responsible to provide air at the right pressure to the combustion process and to the turbine. The remaining one-third of the turbine power, is used to drive an electricity generator or another mechanical device. With this configuration, the efficiency is only about 33% of the fuel input, as all the exhaust gases heat leaving the turbine would be wasted (simple cycle). However, the heat of the hot exhaust gases can be used to generate steam in a device called a heat recovery steam generator (HRSG). The steam can be used to drive a steam turbine (combine-cycle), to supplement the electricity generation or a mechanical drive, such as to drive the crunchers in the sugar-mill industry. In co-generation the steam is used as a heat provider in industrial processes like cooking or drying or for mechanical energy and process steam. The co-generation cycles can reach efficiencies as high as 60% compared with 20-30% efficiency range of sugar-mill boilers. Taking into account the gasifier efficiency, the predicted overall efficiency of the integrated gasification combined-cycle (IGCC) would be around 50%. Figure 4 illustrates a ço-generation cycle. Many gas turbines are available from manufacturers around the world (Turbomachinery,

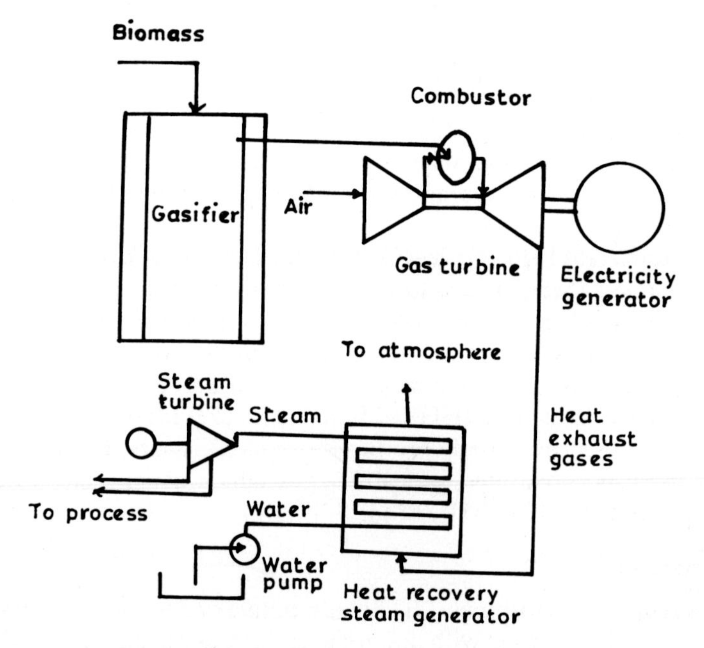

Figure 4. Integrated gasifier co-generation system

1995). One might anticipate that suitable IHGs will be available before the turn of the century.

Liquefaction and Chemical Production

For competitive reasons it might be necessary to maximize the use of the co-gasifier so as to extract all potential economic benefits from it. Table 1 lists the $y = H/C$ and $x = O/C$ coordinates of important pure hydrocarbons and singly oxygenated hydrocarbons up to $y = 4$ that could be valuable by-products of solid fuel conversion. The higher heating values and boiling points are given. Apart from the aromatic compounds all the non-oxygenated compounds are gases at standard temperature and pressure. While these gases can be stored in pressure vessels, oxygenation can convert them to liquids that are more convenient as transportation fuels since they require less volume to store in non-pressurized tanks (Green, 1991). However, as is obvious, oxygenation lowers the heating values. From Fig. 1A it should be clear that the biomass components are closer to the oxygenates represented in Table 1. Hence one might anticipate that it would be easier to make these valuable compounds from biomass rather than from coal.

Medium heat value gas has been used for the production of methanol which can be used with modified engines as a liquid automative fuel to replace gasoline. Liquid alternatives to gasoline, if produced from biomass at competitive prices have great potential for increasing biomass energy use and reducing the use of oil. Derivative fuels from methanol like dimetyl ether are also potential transportation fuels (Green *et al.*, 1989).

The production of ethylene as a feedstock for the chemical industry, particularly the plastics industry, is another promising area for coal-biomass cogasification (Green *et al.*, 1995b). Ethylene, commands about 5 times the market value of methane, the major NG constituent. Thus the use of the cogasifier during low load or no load periods of a GT system to produce valuable storable chemical feedstocks affords an economic opportunity even when NG prices are low. Indeed adding NG to the pot (gasifier) for value enhancement shoud appeal to the NG suppliers.

Conclusions on Optimum Use of Biomass

As a low heat value and low density solid fuel, biomass has been neglected in favor of coal and oil in many industrial countries. However, cultivated or waste domestic biomass can be co-fired profitably in existing coal combustors after relatively minor retorfitting costs. Our recent studies strongly suggest that blending oxygenated fuels such as biomass, municipal solid waste, municipal sewage sludge, tyres, etc., with highly carbonaceous solid fuels can simplify the task of an indirectly heated gasifier. Such simplication should lower the capital and operating cost of the co-gasifier system and provide greater flexibility in feedstock choices. It would appear that a fuel with the H/C and O/C parameters in the neighborhood of peat would be the design goal. Such a feedstock would be easily volatilized yet would have enough fixed carbon to carbonize the pyrolysis H_2O and CO_2 and provide, by combustion, the heat needed for the gasification process.

Since methane is the major constituent of natural gas, co-firing natural gas in small proportions with coal or in boilers designed for oil appeared as a sensible solution to the 1980 oil crises (Green, 1981). Unfortunately, the prevailing wisdom in the USA NG industry at the time was that the natural gas reserves were so low that R&D should be directed toward using coal to make methane despite the multiple processes involved (Probstein and Hicks, 1982). Today the prevailing view in the US NG industry appears to be that natural gas is so abundant the R&D should be applied use of NG for electricity generation and gasification of solid fuels is purposeless. The CCTL view is that the US should encourage greater use of abundant natural gas in premium fuel applications where coal and biomass cannot be used lest the important coal based-utility industry and the rapidly developing bioenergy industry be disrupted. In addition to expanded residential and new transportation applications, the use of NG in co-firing or co-gasifying applications and making valuable chemicals would be the highest use of this premium fuel. Some aspects of these issues may apply in India and other countries in the future. Some Middle East petroleum countries have purchased vessels for transporting bulk quantities of liquefied natural gas to other countries. Accordingly natural gas might become available in India at reasonable prices and its potential use for upgrading lower rank energy sources in co-firing, co-gasification or co-conversion should not be neglected.

Co-gasification of biomass in indirectly heated gasifiers is our recommended way to use the latest factory fabricated combustion turbines in future distributed electrical generating installations (co-generation or combined cycle). Complementary feedstock could be used to simulate peat or whatever H/C and O/C feedstock is optimum. When NG is available at low costs, blending with an appropriate catalyst could extend the range of carbonaceous materials that could be gasified. Selected components of MSW should also be combined with waste and cultivated biomass in many communities to provide the critical year round fuel base to support a new system. Past emission problems can be overcome by using pollution prevention (Green, 1992), post-gasifier hot gas clean-up and modern post-combustor pollution control equipment. The indirectly heated co-gasifier approach lends itself to potential modular system development at scales that we estimate to be in the 1 to 10 megawatt thermal range. Such systems could render valuable service in India and in parts of the USA remote from natural gas pipelines.

Despite the fact that the thermal use of biomass is humankind's oldest technology there are an amazing number of serious uncertainties in the underlying sciences upon which this technology must be based. Cooperative R&D programs, unimpeded by special approaches to energy problems that prevail in many countries, could be the best path to resolve these uncertainties. Then thermal applications of biomass and other hydrocarbon fuels could be based upon applied sciences capable of reliable predictions and plans to go from concept to practical systems can be carried out with much greater confidence.

REFERENCES

Albright, L., B. Crynes and Nowak, S. eds. (1992). *Novel Production Methods for Ethylene, Light Hydrocarbons and Aromatics.* Marcel Dekker, Inc. New York, NY.

Babcock and Wilcox Inc. (1992). *Steam* Brberton Ohio.

Baxter, L.L., et. al. (1996). Alkali Deposits Found in Biomass Boilers. The Behavior of Inorganic Material in Biomass–Fired Power Boilers–Field and Laboratory Experiences, Sandia National Laboratory, NREL/TP 433-8142, Vol. II.

Borenson, M.L., Howard, J.B., Longwell, J.P., Peters, W.A. (1989). Product Yields and Kinetics from the Vapor Phase Cracking of Wood Pyrolysis Tars. *ALChE Journal.* 35, 1.

Brightstar Synfuels, (1996). Executive Summary, private communication B. Smith with A. Green. Combustion Engineering *Combustion* Windsor CT, 1980.

DOE U.S. (1995). Clean Coal Technology Demonstration Program, (1994). DOE/FE-0330 Wash. DC.

Eoff, K., Shaw, L. and Post, D. (1983). "Biomass Gasification as a Source of Acid Substances" in Acid Deposition Cause and Effects, A. Green and W. Smith, eds, Government Institutes Rockville Md. pp. 153–161.

Florida Power Corporation, Final Report, Wood Gasifier Program, September, 1995 Gaur, S. and Reed, T.B. (1995). "An Atlas of Thermal Data for Biomass and Other Fuels". Report NREL/TP-433–7965, National Renewable Energy Laboratory, Golden CO.

Green, A., 1955. *Nuclear Physics,* McGraw Hill, New York, NY.

Green, A., Pamidimukkala, K.M. (1982). Kinetic Simulation of the Combustion of Gas/Coal and Coal/Water Mixtures, *Proc. 1982 Inter. Conf. on the Combined Combustion of Coal and Gas,* Cleveland, OH.

Green, A., Pamidimukkala, K. (1984). Synergistic Combustion of Coal with Natural Gas, Energy, Vol. 9, pp. 477–484.

Green, A. (1986). Proposal for a Gasification Studies Institute in conjunction with U. Fla Cogeneration Plant.

Green, A. (1993). CoCombustion 93, *Proc. ASME IJPGC.,* FACT Vol. 17, New York, NY.

Green, A., Ed. (1980). *Coal Burning Issues,* Univ. Presses of Florida, Gainesville, FL, pp. 1–380.

Green, A., Ed. (1981). *Alternative to Oil, Burning Coal with Gas,* Univ. Press of Fla Gainesville, pp. 1–140.

Green, A., and Smith, W., Eds. (1983). *Acid Deposition Causes and Effects,* Government Institutes Rockville, MD.

Green, A., Meyreddy, R.R. and Pamidi, K.M. (1984). A Molecular Model of Coal Pyrolysis, *Inter. J. of Quantum Chemistry,* 18, pp. 589.

Green, A., Wagner J., *et al.* (1986). Coal-Water-Gas, An All American Fuel for Oil Boilers, *Proc. of the Eleventh International Conference on Slurry Technology,* Hilton Head, SC.

Green, A., Ed. (1988). *Cocombustion,* ASME-FACT, Vol. 4, New York, NY.

Green, A., Rockwood, D. and Prine, G. (1989). "Co-combustion of Waste, Biomass and Natural Gas," Biomass 20, 249–262.

Green, A., Ed. (1989). *Greenhouses Mitigation* ASME-FACT, Vol. 7, New York, NY.

Green, A. and W. Lear, Jr., Eds. (1990). *Advances in Solid Fuels Technologies ASME-FACT,* Vol. 9, Book No. G00522.

Green, A., Ed. (1991). *Solid Fuel Conversion for the Transportation Sector,* ASME-FACT, Vol. 12.

Green, A., Ed. (1992). *Medical Waste Incineration and Pollution Prevention,* Van Nostrand Reinhold. New York, NY.

Green, A. and Chemyy, V.V., Eds. (1995). *Defence Conversion A Critical East West Experiment,* A. Deepak Publishing Co., Hampton VA.

Green, A., Peres, S., Mullin, J. and Xue, H. (1995). Cogasification of Domestic Fuels. *Proceedings of IJPGC,* Vol. 1, Minneapolis, MN. ASME-FACT New York, NY.

Green, A., Zanardi, M., Peres S., Mullin J. (1996a). Cogasification of Coal and other Domestic Fuels. *Proc, 21st Inter. Conf. on Coal Utilization,* Clearwater FL, pp. 569–580.

Green, A., Zanardi, M., Peres, S., and Mullin, J. (1996b). "Cogasifying Biomass with other Domestic Fuels", *Bionergy 96,* Nashville TN.

Green, A., Zanardi, M., Jurczyk, K., Peres, S., and Mullin, J., (1996c), "Cogasifying Solid Fuels" ASME-IJPGC, Houston TX, in press.

Ledford, C. (1992). US Patent 5,095,040.

Mansour, M., Voelker, G., and Durai Swamy, K. (1995). MTCI/Thermochem Reforming Process for Solid Fuels for Combined Cycle Power Generation Vol. 1. Proc. of IJPGC ASME Minnaepolis MN 1995.

Nunn, T., Howard, J., Longwell, J. and Peters, W. (1985). Product Compositions and Kinetics in the Rapid Pyrolysis of Sweet Gum Hardwood. *Amer. Chem. Soc. Ind. Eng. Chem. Process Des. Dev.* 24, 836–844.

Prine, G., and Woodward, K., (1994). "Leucaena and Tall Grasses as Energy Crops in Humid Lower. South," USA, Proc Sixth National Bioenergy Conference, Reno/Nevada, October.

Probstein, R.F. and Hicks, R.F. (1982). *Synthetic Fuels,* McGraw Hill Book Company, New York, NY.

Reed, T.B., Ed., (1981). *Biomass Gasification Principles and Technology,* Noyes Data Corp., Park Ridge, NJ.

Rockwood, D., Patbok, N., and Satapathy, P. "Woody Biomass Production Systems for Florida," (1993). Biomass and Bioenergy 5 (1): 23–24.

Smith Albeert (1996). Executive Summary the Brightstar Synfuel Co. process (private communication with A. Green).

Smith, W.H. and Frank, J.R., Ed. (1988). *Methane from Biomass, A Systems Approach,* Elsevier Applied Science, New York, NY and London, England.

Sterzinger, G. (1995). Making Biomass Energy a Contender "Technology Review MIT Cambridge MA, Oct. pp. 35–40

Turbomachinery International Handbook, 1995, Vol. 36 No. 4, Norwalk, CT.

West, C., (1986). Principles and Applications of Stirling Engines, Van Nostrand Reinhold, NY.

7

Woody Biomass Production

Donald L. Rockwood

INTRODUCTION

India's energy needs and an estimated land base of 60 million hectares encourage developing biomass energy sources (Ramamurthi, 1996). Because of its importance in rural areas, renewability, and potential for increased production while conserving soil and water, woody biomass is a promising energy source.

Woody biomass production can be maximized in short rotation intensive culture (SRIC) systems. SRIC systems can best be developed when supported by a comprehensive research program. Elements of such a research program are examined from the perspective of nearly two decades of SRIC research, primarily in Florida, USA. (Rockwood *et al.*, 1985, 1993; Rockwood and Prine, 1988) and various research details are reviewed.

A SRIC research program should address a number of silvicultural and economic issues. These include genetic testing and improvement, propagation, site preparation/amendment, vegetation control, planting density, coppice management, harvesting options, economics of all phases of the system, biomass properties, and end products.

GENETIC IMPROVEMENT

Woody biomass productivity depends heavily on characteristics of the species that are grown. The ideal woody species would have rapid early growth, tolerance of climates and sites, responsiveness to site amendment, tolerance of planting density and vegetative competition, and coppicing ability. It should also have a small crown, make efficient use of nutrients and water, be vegetatively propagated, and have a short generation time but be noninvasive. Another favorable attribute would be suitability for multiple products.

Relatively few tree species possess all these characteristics. Of native species, typically only light-intolerant pioneer species fulfill the most important characteristics. Promising exotic species often are limited by climate and site.

SPECIES SELECTION

Considerable number of species have been identified as having potential for SRIC systems (e.g., NAS 1980 and 1983). In the absence of native species with necessary SRIC characteristics, careful testing of exotic species is essential. In regard to preliminary selection of species, climatic/edaphic matching of candidates is helpful. For Australian species, considerable information is available and can be accessed quickly through a computer software (Booth, 1991).

Species screening studies should be conducted over the range of climates and sites on which SRIC will be practiced and should adequately represent each species. Climate and site definitions should reflect major land bases available for SRIC; as many as three studies should be established on each climate/site type. The species in each study should only be those with reasonable likelihood for success, and each species should be represented by 10 or more well matched genotypes. For unbiased estimation of species potential, all competing vegetation should be eliminated from each study.

Some 800 seed lots and more than 350 clones of *Eucalyptus* have been evaluated for some nine site/climate combinations in Florida (Rockwood and DeValerio, 1986; Reddy and Rockwood, 1989). In the absence of severe freezes *E. grandis* is the most productive species for southern Florida. *E. camaldulensis* and *E. tereticornis* have promise as frost-hardy, rapidly growing species in south-central Florida. *E. camaldulensis* is the best species for sand ridges; *E. tereticornis* may be more suitable for wetter sites. *E. amplifolia* is the most frost-resilient eucalyptus and is thus the best choice for northern Florida. Other promising species were also identified in species screening studies (Rockwood and DeValerio, 1986, Rockwood and Geary, 1991).

The Biotechnology Centre for Tree Improvement at Tirupati has identified promising species for Andhra Pradesh. For appropriate climatic and edaphic zones, these include *Azadirachta indica, Casuarina* species, *Dalbergia latifolia, Eucalyptus* species, *Pterocarpus santalinus, Tamarindus indica* and *Tectona grandis* (Ramamurthi, 1996).

SPECIES IMPROVEMENT

Once a species has proven suitability for SRIC systems, an applied tree improvement program can be implemented to further increase productivity. Depending on the intensity of the program, increments in productivity should be at least 15% per generation. Increments per generation are likely to be higher for exotic than native species.

Numerous tree improvement strategies may be followed. The actual strategy employed for a species will reflect factors such as commercial importance of the species; financial, land and personnel resources available; and technical expertise. Regardless of strategy, a genetic base population of 300 unrelated individuals should be assembled for each species to support long-term genetic improvement.

A low-cost yet effective seedling seed orchard strategy is described by White and Rockwood (1994). The strategy combines genetic infusion, provenance evaluation, progeny testing, and orchard development at one site at one time. Through pedigree control, relatedness is minimized over generations, which can be as short as the flowering time, or four years for many *Eucalyptus* species. The base population in each generation also can be a source of cloning candidates.

This strategy applied to *E. grandis* in Florida predicted that seedlings from all trees in a fourth-generation seedling seed orchard would have 54% greater coppice stem volume; seed collected from only the best tree in each of the top 50 progenies were expected to have 179% more volume (Reddy and Rockwood, 1989). Tree size and freeze resilience will continue to improve through more advanced-generation seed orchards. However, clonal selection with testing is the best short-term alternative for increasing tree size and developing sufficient freeze-resilience (Rockwood and Meskimen, 1991). Superior clones occurred four times more frequently in advanced-generations than in new introductions (Meskimen *et al.*, 1987).

PROPAGATION

Based on a minimum of three genetic tests, best performing genotypes can be identified for propagation. Propagation methods for commercial planting stock will balance genetic gain potential against cost of propagation.

To illustrate, many recommended *Eucalyptus* clones can be propagated as seedlings, rooted cuttings, or plantlets. Vegetative propagules of e.g., *E. grandis* e.g. are faster growing, more uniform in size and shape, and more frost-resilient than seedlings (Rockwood and Warrag, 1994). However, plantlets are more expensive than rooted cuttings, which in turn cost more than seedlings (currently about $ 1.00, 0.50, and 0.25 each, respectively).

Genetic Test Designs

Designs for genetic tests using block plots or row plots for genetic entries are commonly employed but can be inefficient. Provenance tests traditionally have assigned provenances to block plots of 25, 49, or even 100 trees. While good for minimizing inter-provenance competition, block plot designs require large amounts of land and inflate environmental variation within replications. Progeny tests typically assign progenies to row plots, which, although easy to install, may confound environmental variation with genetic differences. Single-tree plots, relative to block land required to differentiate among provenances and progenies (Rockwood and Meskimen, 1991). While single-tree plot designs are more complicated to install and analyze, their great efficiency compels their wider use in genetic testing.

When freeze hardiness evaluation is a requisite part of genetic testing, artificial freeze techniques may be used to decrease evaluation time and to increase the reliability of evaluation. The techniques can be applied to young greenhouse–grown seedlings and extrapolated to field performance. As an example, the best *E. camaldulensis* clone after two field freezes (Rockwood and De Valerio, 1986) also was more frost hardy than *E. grandis* clones in a laboratory freeze test (Rockwood *et al.*, 1989). Genetic gain tests should be incorporated in a tree improvement program. Gain tests should be established beginning with the first-generation of improved trees and repeated with each new generation. Baseline, unimproved trees of the species should be included in each test. For reliable estimates of per hectare productivity gains, block plots can be used. The tests should be repeated over all site types on which the improved trees will be used. Genetic gain tests are extremely valuable for demonstration purposes and for documenting research progress.

Culture

Silvicultural treatments, especially fertilization and planting density, can greatly affect SRIC productivity, as evidenced in slash pine (Neary *et al.*, 1990, Table 1). Fertilizing slash pine at planting on good sites resulted in no growth increases after 10 years. On typical sites, though, responses to initial fertilization were much greater, especially to a high level of sewage sludge. However, this high one-time application did not increase yield to the level of unfertilized growth on the good site.

The effectivness of more intensive culture was evident (Table 1). Annual maximal fertilization or complete weed control resulted in more than three-fold increases in productivity over the control culture of conventional bedding without fertilization and weed control. Fertilization and weed control combined gave a nearly five-fold increase compared to the control. Intensively cultured trees were 4 m taller than unfertilized trees on good sites. No fertilization with a density of 10,000 trees ha^{-1} on a good site yielded slightly less biomass than intensive culture of 1,518 trees ha^{-1} on typical sites.

High planting densities had relatively lower survival and diameter at breast height (DBH) after eight years (Rockwood and Dippon, 1989).

Table 1. Tenth-year height (H, in m), DBH (in cm), and stemwood dry weight[1] (DW, in Mg ·ha^{-1} yr^{1}) for slash pine in fertilizer and planting density treatments on typical and good sites and in intensive culture treatments on a good site (Rockwood *et al.*, 1993).

	Typical Sites			Good Sites		
Treatment	H	DBH	DW	H	DBH	DW
Fertilizer[2]						
0	7.2ab[3]	6.5ab	3.5ab	9.4a	7.6a	6.7a
50/50	6.3b	5.6b	2.0c	9.8a	8.0a	7.4a
150/50	7.9ab	6.1b	3.5ab	10.5a	8.5a	8.2a
175/135S	7.2ab	6.0b	3.0bc	–	–	–
200/100	8.0ab	6.4b	3.8ab	9.8a	7.8a	6.8a
350/265S	9.0a	7.6a	5.7a	–	–	–
Density[4]						
4,800	8.1a	8.1a	3.1b	9.9a	10.5a	6.1b
8,400	7.5ab	6.7b	3.4ab	10.1a	8.7b	7.4a
14,600	6.9abc	5.7c	3.7a	8.8b	7.1c	7.3a
25,100	6.2bc	4.7d	3.8a	8.2bc	5.7d	6.9ab
43,300	5.7c	3.9e	3.8a	7.2c	4.9e	6.0b
Culture[5]						
Control				8.9c	10.1c	1.6c
Fertilizer				13.0b	15.4b	5.8b
Herbicide				12.5b	14.8b	5.2b
F + H				13.5a	17.3b	7.6a

[1] Main stem without bark.
[2] N/P in kg ha^{-1}; S=sewage sludge; initial planting density=10,000 trees ha^{-1}.
[3] Means within a trait/test not sharing the same letter are significantly different at the 5 per cent level.
[4] Trees ha^{-1}.
[5] Initial planting density=1,518 trees ha^{-1}.

Through 10 years, survival was significantly less only at the 43,000 trees ha^{-1} density, and the higher densities produced shorter trees with smaller DBHs (Table 1).

Nevertheless, mean annual production ha^{-1} at the greater densities still equaled or exceeded that at lower densities.

Planting density influences at younger ages typically follow different trends, as evidenced in *E. grandis* (Table 2). Although lower densities resulted in larger tree sizes from 17 to 33 months, similar survivals across densities gave higher productivities to higher densities. Thus for very short rotations, high planting densities are necessary for high yields.

Planting density affected the tree size of *E. amplifolia* but not yield (Table 3). Tree DBH was nearly twice as large at the 10,000 trees ha^{-1} density as at the higher densities, while survival percentages were similar across the three densities. Yield ha^{-1} was not significantly influenced by planting density; the 10,000 trees ha^{-1} density averaged approximately 10 Mg ha^{-1} yr^{-1} after 4.5 years. For rotations of 4.5 years or more, a planting

Table 2. Planting density influence on height, DBH, stemwood volume, and copping of *E. grandis* (Rockwood *et al.*, 1982).

	Density (trees/ha)				
Trait-Age	4,800	8,400	14,600	25,100	43,300
Height (m)					
– 8 mo	1.5	1.5	1.6	1.9	2.1
– 17 mo	4.8	4.2	4.3	4.5	4.5
– 33 mo	7.5	6.4	6.1	5.9	5.8
DBH (cm)					
– 8 mo	2.5	2.5	2.3	2.2	2.2
– 17 mo	3.8	2.8	2.5	2.5	2.3
– 33 mo	6.0	4.3	3.6	3.3	2.9
Volume (m^3/ha/year)					
– 8 mo	2.1	3.8	6.0	11.1	18.5
– 17 mo	6.8	6.9	10.2	15.3	20.8
– 33 mo	13.9	11.2	17.6	21.1	25.0
Coppicing (%)					
– 12 (45) mo	80	87	83	83	83

Table 3. Planting density influence on survival, DBH, and dry weight of five 53-month-old *E. amplifolia* provenances (Rockwood *et al.*, 1993)

	Planting Density (trees/ha)		
Trait	10,000	20,000	40,000
Survival	60a[1]	64a	64a
DBH	5.6a	3.0b	3.2b
Dry Wt.	45a	20a	34a

[1]Means not sharing the same letter within a trait differ at the 5% level

density of 10,000 trees ha^{-1} is advantageous because individual trees are larger and establishment costs are less.

COPPICING

Successful, vigorous coppicing is another necessary attribute of a SRIC species, especially one that is difficult and/or expensive to propagate. To maintain stand productivity, a high percentage of harvested trees need to coppice. Ideally, coppicing will occur throughout the year so that harvests will not be restricted to particular seasons. Self-pruning of coppice shoots is also desirable, resulting in only a few dominant stems well-attached to the stump. Coppice growth should exceed original stem growth by at least 20–25 per cent. Multiple coppice rotations are essential to help offset high establishment costs of some SRIC species.

E. amplifolia exemplifies desirable coppicing ability in a SRIC species (Table 4). The vigour of a fifth coppice was similar to earlier coppices, and stool survival remained nearly constant. Growth in each coppice rotation exceeded seedling growth. A high percentage of harvested trees coppiced, and coppice growth was rapid and well-formed with mostly one stem per stool becoming dominant within three months. As nearly 90% of the trees felled in the summer coppiced (Rockwood *et al.*, 1991), trees will apparently coppice regardless of season of harvest, unlike the seasonal fluctuations in *E. grandis* coppicing (Webley *et al.*, 1986).

Table 4. Height, DBH, survival, and dry weight yield of seedling and seedling and coppice rotations of *Eucalyptus amplifolia*.

Rotation	Stem Age (mo)	Height (m)	DBH (cm)	Surv. (%)	Yield (Mg $ha^{-1}yr^{-1}$)
Seedling	8	1.6	–	81	–
First coppice	23	7.9	8.2	76	23.1
Second coppice	12	3.7	2.0	69	1.2
Third coppice	12	3.8	2.6	69	2.0
Fourth coppice	22	6.8	4.4	68	–
Fifth coppice	6	4.1	2.0	65	–
Sixth coppice	5	–	–	65	–

Coppicing an established plantation may not be superior to replanting, however. If coppice survival is poor or coppice growth is slow, a coppiced stand's productivity will be less than that of a new plantation. Replanting is especially preferred if faster-growing, better-coppicing planting stock is available. For many SRIC species such as *Eucalyptus*, time to flowering is so short that trees can often be bred and tested to develop superior planting stock within the span of two rotations. When the growth of new planting stock is 25% more than what was originally used to establish a plantation, replanting is probably justified.

CULTURE TEST DESIGNS

Silvicultural tests, especially those involving site amendment and planting density treatments, are typically very large when traditional block plots are used. Hall (1994) has proposed a less space demanding design that has promise for jointly identifying SRIC planting densities and clones, and even the best combination of the two for particular products. Data from a Nelder's design with supplemental block plots are used to assess crown competition due to density and genotype. From crown development patterns, planting densities and rotation lengths necessary for each clone to yield a particular product can theoretically be estimated very efficiently.

Sufficient borders around silvicultural tests and their component plots are critical to accurate discrimination among amendment and density treatments. A general recommendation of planting border rows out to

1/4 of tree height at the last measurement could need to be increased in the case of fertilizer treatments, for example. Adequate bordering is especially important for yield tests so that yields are not overestimated.

ECONOMICS

Economic assessments are useful in evaluation SRIC system activities. The more costly components represent areas for research emphasis.

Site preparation and planting can be major components of woody biomass production costs (Rockwood and Dippon, 1989). They comprised nearly 85% of the costs of a 10,000 tree ha^{-1} *E. grandis* plantation managed on two-year rotations Coppice productivity and discount rate assumptions did not affect the choice of the optimal rotation. In general, total production increased with time period, and for any time period higher planting densities produced more biomass. For slash pine in SRIC systems, biomass had an average cost Mg^{-1} of $40.32 in an eight-year rotation, of $2.45 GJ^{-1}. The discount factor affected the estimated average costs of production, but the estimates were more sensitive to planting densities than to the discount factor. The optimal rotation length is affected by both the alternative rate of return for capital and the initial planting density.

BIOMASS PROPERTIES

The quality of woody biomass produced in SRIC systems is important to the utilization of the biomass. Relative partitioning of the biomass into foliage, branches, stemwood, and stembark components by the trees may define product possibilities. Stem quality characteristics such as branch number, branch size, and stem straightness determine potential for trees to quality for high value solid wood products. Wood and bark properties also may be critical.

Basic stem biomass properties vary among and within species and with age. For example, the wood density of 53-month-old *E. amplifolia* varied among provenances, and on average, was slightly greater than that of 13- to 103-month-old *E. grandis* but less than the density of older *E. grandis* of either seedling or coppice origin (Table 5). Wang *et al.* (1984) predicted changes of 4 to 7% for wood density and moisture content, respectively, for *E. grandis*, due to selection. While most basic stem wood properties are determined by felling trees, non-destructive sampling methods have promise for quickly assessing wood density in *Eucalyptus* (Rockwood *et al.*, 1995).

Biomass properties determine possible alternative end uses of SRIC trees. Considerable variation exists among species in terms of suitability for alcohol production, for example, and variation within species for wood

Table 5. Stemwood and bark specific gravity (SG) and moisture content (MC) of three *E. amplifolia* provenances and *E. grandis* of seedling or *coppice origin* (Rockwood *et al.* 1993, Rockwood *et al.* 1995)

Species-Origin	Age (mo)	No. of Trees	Wood SG (g/cm³)	Wood MC (%)	Bark MC (%)
E. amplifolia - Provenance:					
2–3	53	3	.532	109	254
2–7	53	3	.433	148	328
2–8	53	3	.440	151	325
Overall		9	.468	136	302
E. grandis - Stem Type:					
Seedling	13	2,648	.372	168	–
Seedling	103	222	.422	131	–
Seedling	206	37	.493	133	–
Coppice	90	222	.418	135	–
Coppice	193	45	.488	136	–

properties can be utilized to further improve convertibility (Rockwood and Squillace, 1981). Variability among woody species cell wall constituents and lignin influence fermentability; eucalyptus are high in both (Rockwood *et al.*, 1993).

SRIC systems have more value and interest if more than just fuelwood can be produced, e.g., pulpwood sawtimber, poles, food. Ideally, SRIC system management could be altered to produce any of a variety of end products as dictated by market conditions, or to produce successive harvests of increasingly valuable products. To achieve such flexibility, multiple product or multipurpose tree species should be favoured for SRIC systems.

REFERENCES

Booth, T.H. (1991). Where in the world? New climatic analysis methods to assist species and provenance selections for trials. *Unasylva* **165:** 51–57.

Frampton, L.J., Jr. and Rockwood D.L. (1983). Genetic variation in traits important for energy utilization of sand and slash pines, Silvae Genetica. **32(1–2):** 18–23.

Hall, R.B. 1994. Use of the crown competition factor concept to select clones and spacings for short-rotation woody crops. *Tree Physiolgy* **14:** 899–909.

Meskimen. G.F., Rockwood, D.L. and Reddy, K.V. (1987). Development of *Eucalyptus* clones for a summer rainfall environment with periodic severe frosts. *New Forests* **3:** 197–205.

National Academy of Sciences. (1980). *Firewood Crops: Shrub and Tree Species for Energy Production.* National Academy of Sciences—National Research Council, Washington, DC. 237 p.

National Academy of Sciences. (1983). *Firewood Crops: Shrub and Tree Species for Energy Production. Volume* 2. National Academy of Sciences–National Research Council, Washington, D.C. 92 p.

Neary, D.G., Rockwood, D.L., Comerford, N.B., Swindel, B.F. and Cooksey, T.E. (1990). Importance of weed control, fertilization, irrigation, and genetics in slash and loblolly pine early growth on poorly-drained spodosols. *For. Ecol. and Man,* **30:** 271–281.

Ramamurthi, R. (1996). Bioenergy, unpublished Report, Sri Venkateswara University, Tirupati, India, 18 p.

Reddy, K.V., and Rockwood, D.L. (1989). Breeding strategies for coppice production in an Eucalyptus grandis base population with four generations of selection. *Silvae Genetica* **38 (3–4):** 148–51.

Rockwood, D.L., Comer, C.W., Dippon, D.R. and Huffman, J.B. (1985). Woody biomass production options for Florida *Fla. Agr. Exp. Sta. Tech. Bull.* 856. 29 p.

Rockwood, D.L., and DeValerio, J.T. (1986). Promising species for woody biomass production in warm-humid environment. *Biomass* 11, 1–17.

Rockwood, D.L., Dinus, R.J., Kramer, J.M. and McDonough, T.J. (1995). Genetic variation in wood, pulping, and paper properties of *Eucalyptus amplifolia* and *E. grandis* grown in Florida USA. In: *Proceedings CRC-IUFRO Conference Eucalypt Plantations: Improving Fibre Yield and Quality,* February 19–24, 1995, Hobart, Australia p. 53–59.

Rockwood, D.L., and Dippon, D.R. (1989). Biological and economic potential of *Eucalyptus grandis* and slash pine as biomass energy crops. *Biomass* **20(3–4):** 155–166.

Rockwood, D.L., and Geary, T.F. (1991). Growth of 19 exotic and two native tree species on organic soils in southern Florida. *Proceedings Symposium on Exotic Pest Plants,* November 2–4, 1988, Miami, F.L. USDI Nat. Park Serv. Rpt. NPS/NREVER/NRTR-91/06, 283–302.

Rockwood, D.L., Geary, T.F. and Bourgeron, P.S. (1982). Planting density and genetic influences on seedling growth and coppicing of eucalypts in southern Florida. *Proc. 1982 South. For. Biomass Working Group Workshop*: 95–102.

Rockwood, D.L., and Meskimen, G.F. (1991). Comparison of *Eucalyptus grandis* provenances and seed orchards in a frost-frequent environment. *South African Forestry Journal* 159: 51–59.

Rockwood, D.L., Pathak, N.N. and Satapathy, P.C. (1993). Woody biomass production systems for Florida *Biomass and Bioenergy* **5(1):** 23–34.

Rockwood, D.L., Pathak, N.N., Satapathy, P.C. and Warrag, E.E. (1991). Genetic improvement of *Eucalyptus amplifolia* for frost-frequent areas. Australian Forestry **54(4):** 212–218.

Rockwood, D.L. and Prine, G.M. (1988). Alternative production systems: Woody crops (Chapter 17). *In Methane from Biomass-Asystematic Approach* (eds. W.H. Smith and J.R. Frank). Elsevier Appl. Sci.—p. 277–289.

Rockwood, D.L. and Squillace, A.E. (1981). Increasing alcohol production from wood by utilizing genetic variation in wood characteristics. Proc. 1981 TAPPI Ann. Meet: 307–316.

Rockwood, D.L., and Warrag, E.I. (1994). Field performance of micropropagated, macropropagated, and seed-derived propagules of three *Eucalyptus grandis* ortets. *Plant Cell Reports* **13:** 628–631.

Rockwood, D.L., Warrag, E.E., Javanshir, K., and Kratz, K. (1989). Genetic improvement of *Eucalyptus grandis* for southern Florida. Proceedings 20th. *Southern Forest Tree Improvement Conference*, p. 403–410.

Wang, S., Littell, R.C., and Rockwood, D.L. (1984). Variation in density and moisture content of wood and bark among twenty *Eucalyptus grandis* progenies. *Wood Sci. Technol.* **18:** 97–100.

Warrag, E.E.I., Lesney, M.S. and Rockwood, D.L. (1990). Micropropagation of field tested superior *Eucalyptus grandis* hybrids. *New Forests* **4:** 67–79.

Warrag, E.E.I., Lesney, M.S. and Rockwood, D.L. (1991). Nodule culture and regeneration of *Eucalyptus grandis* hybrids *Plant Cell Reports* **9:** 586–589.

Webley, O.J., Geary, T.F., Rockwood, D.L., Comer, C.W. and Meskimen, G.F. (1986). Seasonal coppicing variation for three eucalypts in southern Florida. Aust. For. Res. **16(3):** 281–290.

White, T.L., and Rockwood, D.L. (1994). A breeding strategy for minor species of *Eucalyptus*. In *Actas Simposio Los Eucaliptos in el Desarrolio Forestal de Chile,* November 24–26, 1993, Pucon, Chile. p. 27–41.

8

Woody Biomass Production, Species, Land Availability, Policy Issues, Scientific and Technology Needs—Assessment

P.M. Swamy and C.V. Naidu

INTRODUCTION

Biomass as a Source of Energy

Biomass represents a large potential energy resource that is renewable on a sustainable basis (Chynoweth *et al.*, 1993). Production of energy from biomass resources became globally popular because of decreasing fossil fuel supplies and increasing demand for energy, a growing awareness of the environmental impact of fossil fuel usage as well as surplus global biomass production (Frank and Smith, 1993). Several research studies (Smith and Frank, 1988; Turick *et al.*, 1991; Smith *et al.*, 1988; Jerger and Chynoweth, 1987; Wise, 1981) have highlighted production of solid, liquid and gaseous fuels from biomass and agricultural waste produce. With the increasing demand for timber and fuelwood there is heavy pressure on forests in the country. If the tree planting is not greatly accelerated, Spears (1978) estimated that by the year 2000, some 250 million people will be without fuelwood for their minimum cooking and heating needs and will be forced to burn dried animal dung and agricultural crop residues, thereby further decreasing the yield of food crops.

The primary productivity in the energy fixed by a plant that supports all life and only green plant has the power to transform the energy of sun into potential energy by photosynthesis (Agrawal and Prakash, 1980). Odum (1957), Ovington (1962), Rodin and Bazilevich (1966), Kira *et al.* (1967),

Misra *et al.* (1967), Sharma (1976) variously estimated the dry matter production in differed edaphic and environmental conditions. Biomass production and distribution of twenty year old teak (*Tectona grandis*) and gamar (*Gmelina arborea*) plantation in Tripura (a state in North-East India) are studied (Negi *et al.*, 1990).

A few studies dealing with perennial plants have considered the importance of biomass (Van Audel and Vera, 1977; Lovett Doust, 1980; Abrahamson and Caswell, 1982; Gray *et al.*, 1982; Saxena and Rama Krishnan, 1984). Plant biomass and net primary production of a mixed banjoak (*Quercus leucotrichophora*) and chir pine (*Pinus roxburghii*) forests are studied (Rana and Singh, 1990). In a forest ecosystem, estimation of dry matter production by tree and tree seedlings and its distribution into the stem wood are of prime importance (Singh *et al.*, 1982). In view of a rising demand for wood, it is imperative that the forest resources be maximized from the land devoted to forestry (Ovington, 1957). Dry matter production is the key function in the ecological and sociological life of plant (Boysen-Jensen, 1932). According to Leith (1968) the determination of dry matter production by the plants always constitutes the basis for further studies in production ecology.

Plants are still the largest energy convertors, as is evident from the following. Net primary production annually through the process of photosynthesis (3×10^{21} J). The amount of power in fossil fuel reserves is 25×10^{21} J and the amount in standing biomass (90% trees) is 30×10^{21} J, which means energy in fossil fuel reserves is slightly less than in total biomass. These figures demonstrate that the massive bioconversion of solar energy by plants through the overall global efficiency is more 0.2%. Further, biomass energy has certain intrinsic advantages, particularly the forest biomass, as it can be converted into solid (charcoal), liquid (ethanol, methanol, liquid hydrocarbon and oil) and gaseous (methane and synthetic gas) fuels. All these fuels can further be converted into four principal usable energy forms, namely steam, mechanical power, electricity and heat, for which the conversion technologies are already available.

Assessment of Fuelwood Problem in India—Availability and Production

The statewise population annual production of fuelwood and requirement and deficit are presented in Table 1. It is clear that while the requirement is over 180 mm^3 the recorded availability is around 13 mm^3, the deficit being the order of 171 mm^3. It was estimated by the National Commission on Agriculture in 1976 that the fuelwood requirement will be in the order of 300 mm^3 by 1996 while the consumption in 1976 was 200 mm^3. Thus, there will be an additional requirement of atleast 100 mm^3. In rural India 80% of energy requirement is made from the traditional sources, mostly fuelwood, the order priority is being cooking (64%), agriculture (22%), village and industries (7%), lighting (4%) and transportation (3%). The overall energy crisis of nearly 670 million people living mostly in rural areas threatens to become more than an energy issue. Additionally, the aggregate raw material requirement of wood (excluding bamboo) for industrial uses is estimated to be of the order of 35.2 mm^3 and 64.5 mm^3 by 1985, respectively under assumptions of the high income growth (National Commission on Agriculture, 1976).

Table 1. Fuelwood production and deficit in the Indian states

States	Population (Million)	Annual production fuelwood (Million Cu.m.)	Fuelwood requirement (Million Cu.m.)	Deficit (Million Cu.m.)
Andhra Pradesh	66.4	0.97	14.1	13.13
Assam	22.3	0.89	5.5	4.61
Bihar	86.4	0.36	18.7	18.34
Gujarat	41.2	0.18	8.0	7.82
Haryana	16.3	0.11	3.4	3.29
Himachal Pradesh	5.1	0.19	1.2	1.01
Jammu & Kashmir	7.7	0.09	1.6	1.51
Karnataka	44.8	1.71	9.8	8.09
Kerala	29.0	0.55	7.1	6.55
Madhya Pradesh	66.1	1.41	14.1	12.69
Maharashtra	78.7	1.40	16.6	15.20
Orissa	31.5	0.61	7.4	6.79
Punjab	20.2	0.04	4.4	4.36
Rajasthan	43.9	0.23	9.1	8.87
Tamil Nadu	55.6	0.29	13.0	12.71
Uttar Pradesh	139.0	2.09	13.0	12.71
West Bengal	67.9	0.53	15.3	14.77
Other States	54.1	0.80	5.6	4.79
INDIA	**876.2**	**12.45**	**183.8**	**171.34**

Source : Adapted from NCA-Report-Production Forestry.

Therefore, not only present forest resources have to be more intensively and rationally managed but also large plantation programme has to be envisaged and undertaken to meet the rising demand of wood both for industrial requirement and fuelwood. While industrial demands can be made good by managing existing forest land resources, the area for energy plantation have to be mobilized from other land resources. It has been estimated by the National Commission on Agriculture that nearly 4.36 million hectare of cultivable area wastelands which are neither under agriculture nor forestry is distributed among 567,000 villages in this country.

Assuming that 50% of the land is of no value or cannot be released, the balance of around 20 million hectare could be made available for energy plantations.

Plantations make it possible to produce the required amount of biomass within the stipulated period. Further areas can be brought under energy plantations progressively using the cyclic planting and harvesting approach so that the fuelwood supply would perpetually keep ahead of demand. It is, therefore imperative that much of the wastelands be brought under tree cover as quickly as possible. Infact, with appropriate management practices it is even possible to produce over 9 tonnes of utilizable biomass per hectare in a four year cycle, which means the land requirement will be considerably reduced. In Andhra Pradesh energy plantation of *Eucalyptus* (11,791 ha.) and *Casuarina* (8,921 ha) have been planted up to 1978-79 under different plantation schemes. During the Sixth Plan period energy plantation of *Eucalyptus* (3,370 ha), *Casuarina* (25,000 ha) and miscellaneous species (19,000 ha) are being raised. In addition under the 'Drought Prone Area Programme' firewood plantations such as *Acacia auriculiformis* (600 ha) *Eucalyptus* (1800 ha) *Acacia niloti̧ca* (1,400 ha) and *Dalbergia sissoo* (100 ha) have been planted 1977-78. These plantations are expected to yield resources of 770,000 tonnes of wood annually in due course. In all these plantations, however, the approach is to provide not only fuelwood but also small timber and to some extent fodder for the local village population. Further, in these plantations the concept of close planting coupled with short rotation is not the basic approach with the result high yield is not anticipated. Using the concept of energy plantations it is possible to attain high yields in sustained manner as was performance records of energy plantations on marginal, submarginal and saline-alkaline soils of various parts of India have shown that a large number of plant species listed in the Table 2, are promising for fuelwood in arid areas with high calorific values with short rotation cycles. During the last four decades intensive researches on these species have been undertaken by various institutions in India. Potential fuelwood tree species like *Leucaena leucocephala, Acacia nilotica, Acacia auriculiformis, Acacia catechu, Prosopis juliflora, Prosopis cineraria, Casuarina equisetifolia, Albizzia lebbeck, Albizzia procera, Melia azadirachta, Azadirachta indica* and *Parkinsonia aculeta*—their uses, calorific values, climatic zones, and adaptability of various fast growing fuelwood tree species are presented in Table 2.

AVAILABILITY OF LAND

Wasteland Resources

The availability of additional land for forestry is constrained for several reasons. Increasing pressure of both human and cattle populations have

Table 2. Potential fuelwood plant species, distribution, uses, calorific values and their adaptability.

Sl. No.	Botanical Name	Climatic Zones	Uses	Calorific Values	Adaptability
1.	*Leucaena leucocephala* (Subabul)	Dry Tropical Moist Subtropical Moist Tropical Parts of Indian Peninsula.	• Beautification • Agroforestry • Ornamental • Shade • Wind Breaker • Erosion control • Fodder • Gum • Green Manure • Medicinal Oil • Pulp and Paper • Timber • Plywood • Boxwood and Tanin	4,200–4,600 Kcal/kg	• Clayey Soils • Ravine Land • Rocky Soils • Saline-Alkaline Soils
2.	*Acacia nilotica* (Babul)	Dry Tropical Parts of Indian Peninsula.	• Food • Fruit • Fence • Fodder • Gum • Medicinal • Pulp • Timber • Tanin	4,800–Sapwood Kcal/kg 4,950–Heartwood Kcal/kg	• Coastal • Clayey Soil • Ravine Land • Saline-Alkaline Soils • Skeleton Soils • Shifting Sand Dunes
3.	*Acacia auriculiformis* (Bengali Babul)	Dry Tropical and Moist Tropical Parts of Indian Peninsula.	• Beautification • Ornamental • Shade • Wind breaker • Fodder • Oil • Tanin • Pulp and Paper	4,800 – 4,900 Kcal/kg	• Coastal • Clayey Soils • Laterite and • Lateritic Soils • Saline-Alkaline Soils • Water logged Soils
4.	*Acacia catechu* (Khair or Cutch)	Dry subtropical, Dry Tropical Moist Subtropical and Mosit Tropical Parts of Indian Peninsula.	• Fodder • Gum • Medicinal • Timber • Tanin	5,142 Sapwood Kcal/kg 5,244 Heartwood Kcal/kg	• Ravine Land • Saline-Alkaline Soils • Skeleton Soils
5.	*Prosopis juliflora* (Mezquit)	Dry Tropical Parts of Indian Peninsula.	• Erosion Control • Wind Breaker • Green Manure • Plywood	4800 Kcal/kg	• Coastal, Ravine Land • Rocky Soil

contd.

Table 2. contd.

Sl. No.	Botanical Name	Climatic Zones	Uses	Calorific Values	Adaptability
			• Boxwood		• Saline-Alkaline Soils • Skeleton Soils • Swifting Sand Dunes
6.	*Prosopis cineraria* (Khejri)	Dry Tropical Parts of Indian Peninsula.	• Agroforestry • Food • Fodder • Gum • Medicinal • Oil • Timber	5000 Kcal/kg	• Rocky Soils • Saline-Alkaline Soils • Swifting Sand Dunes
7.	*Casuarina equisetifolia* (Saru)	Moist Tropical Wet Tropical Parts of Indian Peninsula.	• Beautification • Ornamental • Shade • Wind Breaker • Medicinal • Paper pulp • Timber • Tanin	4299-460 Kcal/kg	• Coastal • Island Region • Rocky Soils • Saline-Alkaline Soils • Water Logged Soils
8.	*Albizzia lebbck* *(Siris)*	Dry Tropical Dry Subtropical Moist Tropical Wet Subtropical Wet Tropical Parts of Indian Peninsula.	• Beautification • Ornamental • Shade • Wind breaker • Erosion control control Fodder Honey Nector • Oil • Tanin • Timber	5163 Kcal/kg	• Saline-Alkaline Soils
9.	*Albizzia procera* (Safed Siris)	Moist Tropical Moist Subtropical Moist Subropical Parts of Indian Peninsula	• Beautification • Fodder • Gum • Medicinal • Paper Pulp • Timber • Tanin	4870-Sapwood Kcal/Kg 4865-Heart wood Kcal/Kg	• Saline-Alkaline Soils • Stream Bank • Water Logged Soils
10.	*Mella azadirachta* (Bakin)	Dry Tropical, Dry Subtropical Moist Tropical Parts of	• Beautification • Ornamental • Wind breaker • Fodder • Gum • Green Manure	5043-5176 Kcal/Kg	• Adaptability to all lands.

		Indian Peninsula	• Medicinal • Oil • Pulp & Paper • Timber • Wax		
11.	*Azadirachta indica* (Neem)	Dry Tropical Dry Subtropical Moist Tropical Parts of Indian Peninsula.	• Beautification • Ornamental • Wind breaker • Fodder • Gum • Green manure • Medicinal • Pulp & Paper • Timber	5070 Kcal/kg	• Coastal • Ravine Land • Rocky Soils • Saline-Alkaline Soils • Skeleton Soils • Shifting Sand Dunes
12.	*Parkinsonia aculeata* (Vilayathi Kikar)	Dry Tropical Dry Subtropical Moist Tropical Parts of Indian Peninsula Moist Subtropical	• Beautification • Ornamental • Shade • Wind breaker • Erosion control • Fibre/rope • Medicinal • Oil • Timber	4800 Kcal/kg	• Coastal • Ravine Land • Shifting Sand Dunes

Data compiled from publications of the Department of Non-Conventional Energy Sources (DNES), Government of India.

already led to encroachments on large chunks of forest area. The 1952 Forest Policy indicated 80 million hectares of pastures and wastelands considered to be available in the country. Since this land was hardly touched for forestry purposes till about 1975 it was obviously put to some other uses by the local communities.

In 1985, the Government of India announced an ambitious plan of afforesting/reforesting some 130 million hectares of 'wasteland' (at the rate of 5 million hectares per year). The break-up of such 'available' and was as follows:

Type of Wasteland	Area (million ha)
Saline and alkaline land	7.17
Wind eroded land	12.93
Water eroded land	73.60
(Sheet erosion, ravines, water-logged, riverain lands gullies etc.)	
Degraded forests	37.40
Total	**131.00**

This was, of course, too simplistic an approach to the problem. Land believed to be thus available for forestry is usually highly degraded, without top soil and even then, most of it would be under encroachment of some type or the other. None of the planners foresaw the difficulty of raising plantations thereon, particularly in view of the normal cost benefit ratio.

Moreover, the most important questions are—Where is the land? Under whose ownership or occupation is it? What is the use to which it is being put at present? It is hard to believe that, in a country with almost 1,000 million humans and 500 million head of cattle, even a single square metre of land is still lying unutilized for some sort of activity or other. It is then no wonder that implementing officials in the various states are starting to view with skepticism any new scheme originating from New Delhi.

Use of Marginal Lands for Woody Biomass Plantation

It has been suggested that there may be large areas of marginal or uncultivated land where biomass plantation for energy purpose could be raised without coming in conflict with food priority, for the use of arable land for non-food purpose in the developing countries should not be considered.

In India *usar* (saline and alkaline soils) lands occupy nearly 7 million hectares of area (Table 3). These lands are spread over in Uttar Pradesh, West Bengal, Rajasthan, Punjab, Maharashtra, Haryana, Orissa, Karnataka, Madhya Pradesh, Andhra Pradesh and other states. Of these nearly 1.3 million hectares exist in Uttar Pradesh alone. These soils are useless for agriculture because of the presence of soluble salts and sodium in excessive concentrations and also present a problem for afforestation. Table 3 shows the extent of saline-alkaline soils in different states in India.

Table 3. Extent of saline and alkaline soils in different states of India (Abrol and Bhumbla, 1971)

State	Area (Hundred thousand ha)
Uttar Pradesh	12.95
Gujarat	12.14
West Bengal	8.50
Rajasthan	7.28
Punjab	6.89
Maharashtra	5.34
Haryana	5.26
Orissa	4.04
Karnataka	4.04
Madhya Pradesh	2.24
Andhra Pradesh	0.42
Delhi	0.16
Kerala	0.16
Bihar	0.04
Tamil Nadu	0.04
Total	69.49 (7 million hectares)

Source: Abrol and Bhumbla, 1971, 41: 42–51.

Therefore, it is assumed that if all the 20 million hectares of available land are brought under energy plantations immediately, it is possible to produce 2000 million tonnes of biomass or 1200 mm^3 of wood which is four times the anticipated requirement of 300 mm^3 by 1990. However, it is neither possible nor feasible to bring all the 20 million ha simultaneously under energy plantations. Even if 25% of the area is brought under energy plantations.

POLICY ISSUES

Every villager, who wishes to take, irrespective of his family alliances and number of individuals present in it, be allotted suitable area of land which is in degraded stage today. It comes about 5 ha per man. This area be given to every person who demands it so as to cover the entire village system. He may be allowed to develop this area as his own with the following basic conditions:

1. The land shall be allotted for a period of 5 years, extendable in periods of 5 years each depending upon the work done in it.
2. He has to show at least 1,000 woody tree species standing per ha area, after the completion of 5 years.

3. He be allowed to cut away the herbage from his area for consumption by his livestock or he may sell it and draw profits.
4. Whatever extra number of trees he is able to grow in the land allotted, be allowed to be harvested by him. He may use it as fuelwood or even sell it if he likes and keep the profit.
5. The most important condition will be that the law may be suitably modified, so that the land area allotted to him is only for the purpose of development of land and no ownership right be conferred on him. In case of his death, the so developed area will not automatically belong to his son, who can only be one of the applicants for allotment. On the part of the Government, the Forest Department has to create enough nursery beds for supplying seedlings to the people at nominal cost or no cost. Those farmers who wish to raise nursery be supplied with seeds, polythene bags, etc., and they may also sell the seedlings at nominal cost.
6. Invitation to multinationals for establishing 100% export oriented industries in India will cause further degradation of land and water and add to energy crises.

SCIENCE AND TECHNOLOGY INPUTS FOR MEETING ENERGY NEEDS

The bioenergy programme has to be developed in consonance with ecological requirements and as a part of rural and human resources development.

- Collection and classification of information on biomass through remote sensing technology, to build up a data base on terrestrial and aquatic ecosystems and sequential changes, including local consumption of fuel wood.
- For biomass production—identification, physiological and pathological studies of fast-growing multipurpose trees suited to specific agroclimates, breeding, tissue culture and use of cell and recombinant DNA technology to evolve energy rich firewood with properties tailored for resistance and adaptability.
- Extension of integrated systems through demonstration, audiovisual narration, training and education accompanied by monitoring feedback.
- For biomass conversion fluidized lead combustion technology, manufacture of paracytic carbon from wood plants as substitute for fossil fuels, development of cost effective immobilized whole cell system, gasification, studies on availability, nature of biocrudes and development of agrotechnology for identified fuelwood species.
- Utilization of aquatic angiospermic biomass.

- Development of efficient methods for utilization of aquatic angiospermic biomass.

For integrated land use, the principle requirements are:

- Sound data base of available land for growing fuelwood plant species
- Suitability and soil capability as related to fuelwood producing species
- Legislative control to prevent misuse
- People's participation in effective land use.

REFERENCES

Abrahamson, W.G. and H. Caswell. (1982). On the comparative allocation of biomass, energy and nutrients in plants. *Ecology*. **63:** 982–991.

Agrawal, P.K. and G. Prakash. (1980). An analysis of dry matter and chlorophyll production in some forest tree seedlings. *J. Indian Bot. Soc.*, **59:** 267–273.

Abrol, I.P. and D.R. Bhumbla. (1971). Saline and alkaline soils in India. Their occurrence and movement. *World Soil Resoures FAO Report*, **41:** 42–51.

Boysen-Jensen, P. (1932). Die stoffproducktion der paflanzen. Fischer Jem. p. 108.

Chynoweth, D.P., C.E. Turick, J.M. Owens, D.E. Jerger and M.W. Peck. (1993). Biochemical methane potential of biomass and waste feed stocks. *Biomass*. **5:** 95–111.

Frank, J.R. and W.H. Smith. (1993). Methane from Biomass—Seince and Technology 1 Feedstock Development. *Biomass*. **5:** 1–2.

Gay, P.E., P.J. Grubb and J.H. Hudson. (1982). Seasonal changes in the concentrations of nitrogen, phosphorus, and potassium and in the density of mycorrihiza in biennial and matrix-forming perennial species of closed chalkland turf. *J. Ecol.*, **70:** 571–593.

Jerger, D.E. and D.P. Chynoweth. (1987). Anaerobic Digestion of Sorghum Biomass. *Biomass*. **14:** 99–113.

Kira, T., H. Ogawa, K. Yoda and K. Ogino. (1967). Comparative ecological studies on three main types of forest vegetation in Thailand. IV. Dry matter production with special reference to the Kao Chong rainforest. *Nature and Life. S.E. Asia*. **5:** 149–174.

Leith, H. (1968). The measurement of calorific values of biological material and the determination of ecological efficiency. Functioning of terrestrial ecosystem at the primary production level. Proceedings of Copenhagen symp. UNESCO. pp. 233–242.

Lovett Doust, J. (1980). Experimental manipulation of patterns of resource allocation in the growth cycle and reproduction of *Smyrnium olusatrum* L. *Biol. J. Linn. Soc.*, **13:** 155–166.

Misra, R., J.S. Singh and K.P. Singh. (1967). Preliminary observations on the production of dry matter by Sal. (*Shorea robusta* Gaertn. f.). *Trop. Ecol.*, **8:** 94–104.

Negi, J.D.S., V.K. Bahuguna and D.C. Sharma. (1990). Biomass production and distribution of nutrients in 20 years old teak (*Tectona grandis*) and gamar (*Gmelina arborea*) plantation in Tripura, India. *Indian Forester*. **110:** 681–686.

Odum, M.T. 1957. Trophic structure and productivity of silver springs *Ecol. Monog.*, **27:** 55–112.

Ovington, J.D. (1957). Dry matter production by *Pinus sylvestris* L. *Ann. Bot.*, **21:** 287–314.

Ovington, J.D. and D.B. Lawrence. (1963). Plant biomass and productivity of prairie, savannal oakland and maize field ecosystems in central, Minnesota. *Ecology*. **44:** 52–63.

Rodin, L.E. and N.I. Bazilevich, (1966). World distribution of plant biomass. Fluctuation of the terrestrial ecosystem at the preprimary production level. UNESCO, Paris, pp. 45–52.

Rana, B.S. and R.P. Singh. (1990). Plant biomass and productivity estimates for Central Himalaya mixed banjoak (*Quercus leucotrichophora*) - chir pine (*Pinus roxburghii* Sarg.) forest. *Indian Forester*. **110:** 220–226.

Saxena, K.G. and P.S. Ramakrishnan. (1984). Growth and patterns of resource allocation in *Eupatorium odoratum* L. in the secondary successional environments following slash and burn agriculture. *Weed Res.*, 24: 127–134.

Sharma, V.K. (1976). Biomass estimation of *Shorea robusta* and *Buchanania lanzan* by regression technique in natural dry deciduous forest. XIV. Int. Cong. IUFRO, pp. 131–142.

Singh, U.N., V.K. Srivastava and R.B.S. Senger. (1982). A comparative study of the rate of dry matter production in forest tree seedlings. *Indian J. For.*, **5:** 14–17.

Smith, P.H., F.M. Bordeaux, M. Goto, A. Shiralipour and A. Wike. 1988. In: *Methane from Biomass: A Systems Approach* (eds. Smith, W.H. and J.R. Frank), Elsevier, London, pp. 291–334.

Smith, W.H. and J.R. Frank. (1988). *Methane from Biomass: A System Approach*, 79–82, Elsevier, London.

Spears, J.S. (1978). Wood as an energy source. The situation in the developing world 103rd Annual Meeting of American Forestry Association, Hot Spring, Arkansas, October 8, 1978.

Turick, C.E., M.W. Peck, D.P. Chynoweth, D.E. Jerger, E.H. White, L. Zsuffa and W.A. Kenney. (1991). *Biores. Technol.*, **37:** 141–147.

Van Audel, J. and F. Vera. (1977). Reproductive allocation in *Senecio sylvaticus* and *Chamaenerion angustifolium* in relation to mineral nutrition. *J. Ecol.*, **65:** 747–758.

Wise D.L. (1981). Fuel Gas Production from Biomass, Vol. 1, CRC press, Boca Raton, Florida, USA.

9

Value-added Chemicals from the Byproducts of Sugar Agro Industry

G. Prabhakar and D.C. Raju

INTRODUCTION

India is one of the largest sugarcane growing countries, producing about 200 million tonnes of cane per annum. Over 400 sugar mills, most of them in cooperative sector, process this cane to make about 11 million tonnes of sugar per year. Sugarcane is one of the best synthesizer of solar energy (about 8%) into biomass and serves as a source of food sugar, fuel bagasse, fibre cellulose and fodder green tops bagasse and molasses. The three main byproducts of sugar industry, namely bagasse, molasses and filter press mud cake, can potentially yield a variety of chemicals but the production of a few products alone is commercially viable. Utilization of these materials will change the economics of the sugar industry, in particular and the nation, in general for the better and also help in the conversion of waste into useful chemicals. This paper lists chemicals that can be obtained from the byproducts sugar agro industry (Purchase, 1983; Mashelkar, 1991; Almazan, 1994).

The sugar industry on the average produces the following amounts of byproducts for every 100 tonnes of sugarcane.

Byproduct	Quantity (tonnes)
Sugar	9.0
Bagasse (50% moisture)	19.5
Molasses	2.6
Filter cake	2.6
Green leaves, tops	22.0

BAGASSE

One of the very useful byproducts, bagasse, has an approximate composition of 46% cellulose, 25% hemicellulose and 20% of lignin. The various products that can be obtained from bagasse are shown in Plate 1.

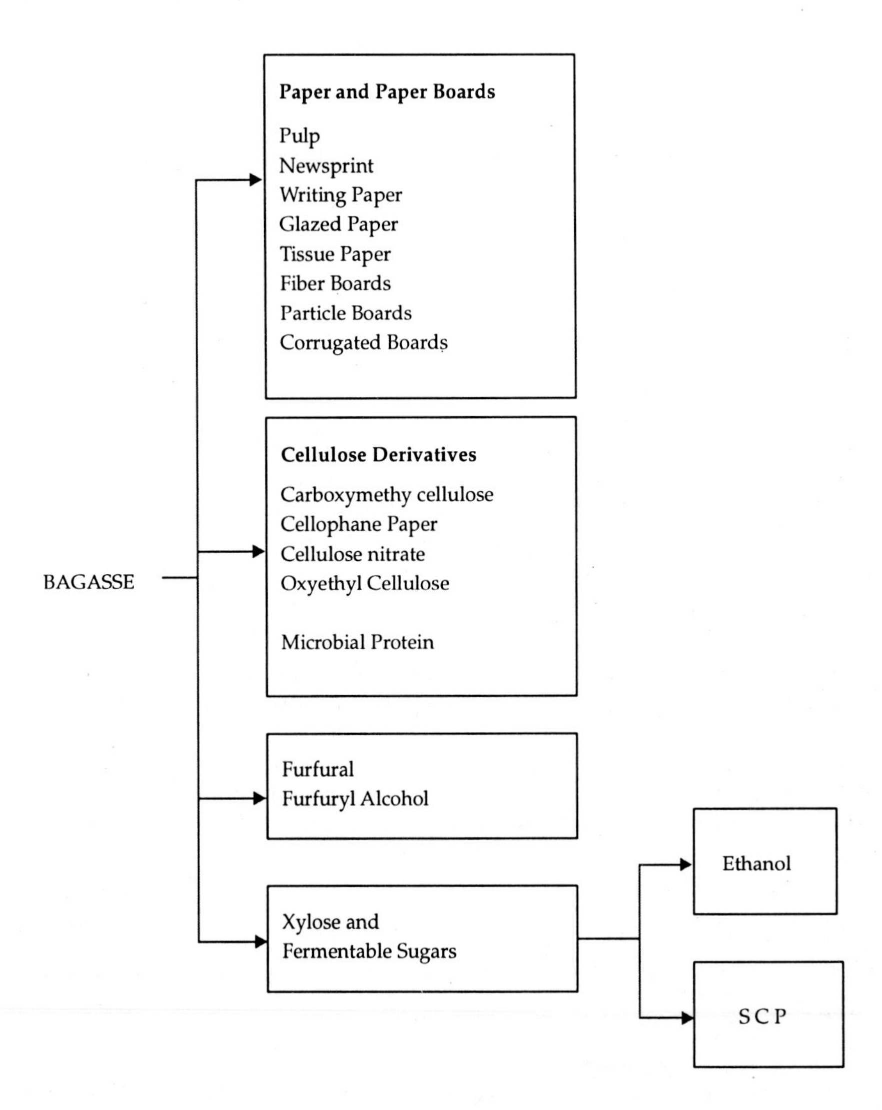

Plate 1.

Pulp, Paper and Paper Board

1. Different grades of paper like newsprint, printing and writing paper, glazed paper, offset paper, tissue paper, fibre and particle boards, corrugated boards, etc., can be made.
2. Of total world production of about 170 million tonnes of paper, over 2 million tonnes are produced from bagasse. This is likely to increase as traditional forest sources are being depleted and also due to public environmental awareness.
3. Mexico is the world's largest producer of bagasse paper. In Asia, bagasse-bamboo mixtures are widely used in paper production.

Cellulose Derivatives

1. A high cellulose dissolving pulp obtained from bagasse is highly useful to obtain several derivatives like carboxymethyl cellulose, cellophane paper, cellulose nitrate, oxyethyl cellulose among others.
2. Microbial protein is reported to be obtained from pretreated bagasse-cellulose.

Furfural

1. A high value basic chemical, of great use as a solvent and as the starting material for plastics, resins and biologically active materials, furfural can be made from bagasse.
2. About 11 plants producing 150,000 metric tonnes of furfural from bagasse are in operation worldwide. This is 50% of the total world production.
3. The United States is a pioneer in this technology. South Africa, China, the Dominician Republic and Cuba are the other countries producing furfural from source.
4. Furfuryl alcohol, obtained by hydrogenating furfural, is useful as binding resin, fuel and solvent for dyes.

Xylose and Other Fermentable Sugars

1. Hydrolysis of bagasse yields predominantly xylose and other fermentable sugars. Studies are underway to use these sugars to produce ethanol and single cell protein.

Activated Carbon

1. Even though the usual sources of activated carbon are sawdust and coconut shells, it is reported that activated carbon is produced from bagasse in Mexico.
2. Viable technology needs to be developed.

MOLASSES

Ethyl Alcohol

1. The oldest and the most widely spread derivative.
2. Brazil is the world's largest alcohol producer, with an annual production of about 125 million hectolitres.
3. Alcohol, besides being an energy source, is a precursor for the production of several different chemicals, as shown in Plate 2. In India alcohol is used as an intermediate raw material for the production of acetaldehyde, acetic acid, acetone, ethylene, butadiene, acrylic acid and its derivatives, and citirc acid and its derivatives.

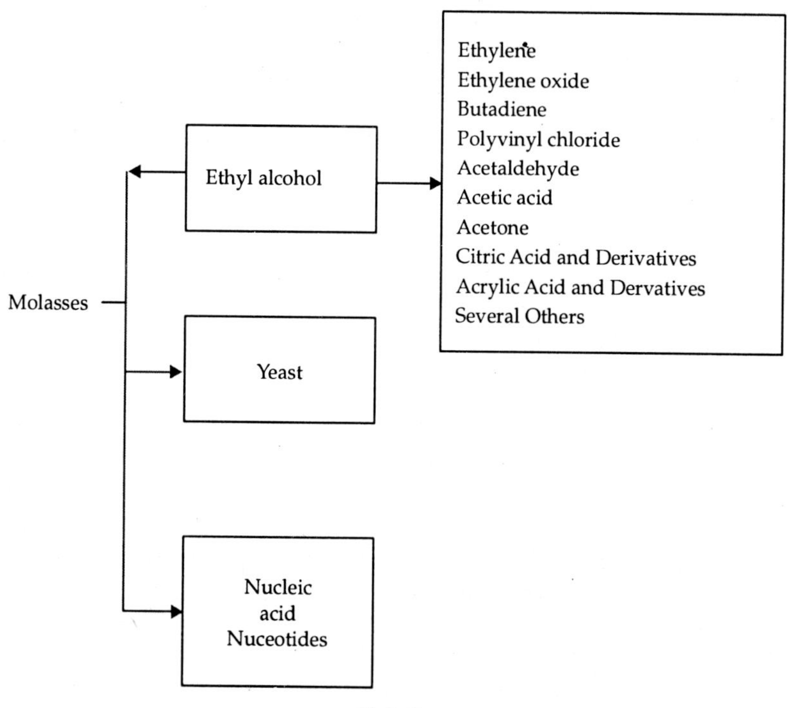

Plate 2.

Yeast

1. Animal feed.
2. A high protein molasses obtained by mixing sugarcane juice and yeast is used in piggery as feed.
3. Being tried as a constituent of human diet.

Nucleic Acids

1. A number of nucleic acids can be obtained from molasses.

2. Nucleotides, obtained from nucleic acid, are high value chemicals and are being increasingly used as flavoring agents.
3. L-lysine can be used for animal and human nutrition.

FILTER PRESS MUD CAKE

It is reported that pharmaceutical chemicals can be produced from this cake. Research studies are in progress.

Wax (Crude and refined)

1. Used as an adhesive.
2. Oil and resins, byproducts of refining wax, are used in animal feed and chemical industry.

CONCLUSION

Various chemicals that can be made from the byproducts of sugar, namely bagasse, molasses and filter press mud cake are listed. Utilization of these sources will definitely ease the pressure on scarce, traditional raw materials. There will be substantial improvement in the profitability of sugar industry.

REFERENCES

Almazan, O. (1994). Past, present and future of sugarcane byproducts in Cuba. Lecture delivered in *VII Joint Convention of STAI & DSTA*, September.

Purchase, B.S. (1983). Perspectives in the production of ethanol from bagasse, *Proc. of the South African Sugar Technologists Association.*

Mashelkar, R.M. in Bharatiya Sugar, 119, Nov. 1991.

10

Composting: An Eco-friendly Technology for Waste Management

Aziz Shiralipour

INTRODUCTION

In 1994, estimates placed the municipal solid waste (MSW) stream at 188 million metric tons annually in the United States (US Environmental Protection Agency, 1994). The projected MSW generation will reach 197 million metric tons by the year 2000 and 228 million by the year 2010 (US Environmental Protection Agency, 1990, 1994). Conventional waste disposal methods are perceived to have serious environmental problems and are becoming too expensive. Landfill disposal costs are ranging from $II per metric tons in Montana to $36 in Florida and Iowa, to $50 in Massachusetts to $75 in Minnesota and $125 in Long Island (Glenn, 1992; Christopher and Asher, 1994). In response, many states, countries, and local governments in the United States have passed legislation to reduce the quantity of materials entering landfills and/or implemented alternative methods of waste management. For example, in 1985, about 83% of MSW was landfilled compared with 72% in 1994 (Christopher and Asher, 1994; Carra and Cossu, 1990). Two-thirds of all landfills in the United States have closed since 1970 with 1000 closing in 1990 and 514 closing in 1991 (Christopher and Asher, 1994). In 1994, a total of 5812 landfills is expected to be reduced to 1200 by the year 2010 (Christopher and Asher, 1994).

Composting of organic wastes is increasingly advocated as an environmentally benign and affordable recycling method. As a result, the number of composting facilities has increased drastically. In 1992, thirty-seven states planned to enact, or had already enacted, legislative bans on landfilling yard waste in the United States. Thirty-five states had solid waste diversion mandates with deadlines between 1991 and 2010 (Glenn,

1992). As a result of legislation banning yard waste from landfills. There has been a dramatic proliferation of yard waste composting facilities throughout the United States. The number of facilities increased from 900 in 1990 to 2200 facilities in 1992 (Glenn, 1992).

As a result of legislation banning yard waste from landfills, there has been a dramatic proliferation of yard waste composting facilities throughout the United States. The number of facilities increased from 900 in 1990 to 2,200 facilities in 1992 (Glenn, 1992). This trend is expected to continue through the 1990s. Kashmanion (1993) estimated that 15.5 million metric tons of yard clippings will be produced in 1996, which after composting should yield up to 7.7 million metric tons of end product.

In the first nationwide survey of sludge composting facilities in 1983, there were 90 projects with 61 operating plants (Richard, 1991). A 1992 survey conducted by Bio Cycle found 290 projects with 159 in operation (Goldstein *et al.*, 1992b).

The number of MSW composting facilities in the United States increased from one in 1985 to seven in 1989 (Goldstein and Riggle, 1989). By 1992, the number of MSW composting projects in various stages of development had increased to 82 (Goldstein *et al.*, 1992).

The increasing number of composting facilities can be attributed to the following reasons.

- Groundwater pollution from landfill leachates have led to more stringent regulations concerning the siting of new landfills. These regulations which govern solid waste land disposal practices, have made it expensive to site new facilities. In addition, the growing public perceptions of landfills as environmentally threatening technologies for solid waste disposal, have forced authorities to consider alternative technologies for solid waste disposal.
- Because of impending air stricter pollution control standards and ash disposal regulations, coupled with the public fears regarding toxic air emissions of dioxins and furans and heavy metal concentrations in the ash, these types of facilities are facing financial strains and stronger public resistance.
- In general, the cost of composting today, on the average, is less than other forms of waste disposal technologies. In some cases, cities have been forced to transport trash hundreds of miles to a landfill, with hauling fees adding greatly to the disposal options such as composting can also reduce energy costs. Private industry has a renewed interest in composting as a business venture.
- There is a growing public awareness of the need to protect the environment and preserve our resources. To some, the commonsense approach to waste disposal is a combination of recycling/composting, reusing and reducing the waste stream along with landfilling and

resource recovery for residual wastes. More state governments are enacting recycling laws and banning certain "reusables" from landfills. Composting is a "reuse" strategy whereby wastes disposed off can be processed into a usable product that has the potential to be more environmentally acceptable.

HISTORICAL PERSPECTIVE AND TECHNOLOGY REVIEW

Composting has played an important role for centuries in rural based societies. It has been used in farming and other agricultural practices. Various organic materials, such as manure, agricultural residues and food wastes were spread on fields to improve the soil characteristics and return the nutrients removed from the soil by crop growth. Today, composting is a viable agricultural practice.

Source separation and reuse were important elements of waste management in the early part of this century in the United States. Organic Food wastes were often separated and used as animal feed. The government promoted separation of wastes during the war times.

Composting has managed to achieve some measure of success as part of the overall waste management strategy in Europe and Asia. It is difficult to compare foreign composting practices as processing technologies may differ. However, composting appears to have a place in the overall waste management strategies of many foreign countries, along with incineration and landfilling.

In the United States, past attempts at composting MSW have met with little economic success. Form the mid 1950s to the middle 1960s many pilot projects were initiated in Arizona, California, Oklahoma, Alabama, Florida, Pennsylvania, and Kentucky. These plants for the most part became casualties of inexpensive landfill space, poorly constructed facilities and little or no demand for the product. This is no longer the case.

Economically, composting of MSW was not a competitive technology. Landfilling, open dumping and incineration disposal options were far cheaper. Most facilities were poorly planned, hastily constructed and encountered serious reduce waste size, was not suitable to the MSW stream, mainly due to inefficient preprocessing to remove bulky items, metals, glass and textiles. Equipment required extensive repair and replacement, which increased the costs of operation. At the time, there were no standards for the production of compost and markets were difficult to develop. Some facilities were able to produce a usable product but were unable to market it.

Ultimately the lessons learned from early attempts at composting MSW were:

- The potential existed for producing a usable product from MSW through composting.

- Major advances in equipment design and operation were necessary to better process MSW and reduce high operating costs associated with frequent breakdowns and intensive maintenance.
- Markets needed to be clearly defined and standards for compost quality were needed to develop and retain such markets.
- Composting of MSW would not be viable until it was economically competitive with other technologies, such as landfilling, incineration and resource recovery.

Today's operating United States MSW composting facilities are, for the most part, employing advanced processing equipment. Market for quality compost are beginning to emerge.

COMPOSTING

Compost is defined as the product resulting from the controlled biological decomposition of organic wastes, that have been sanitized and stabilized to a degree which is potentially beneficial to plant growth when used as a soil amendment; compost is largely decomposed organic material and is in the process of humification or curing (US Composting Council's Standards Committee).

Processing Technologies

Compost processing technologies cover a broad spectrum. On the low end, operations employ windrows and use equipment such as front-end loaders for aerating during composting. On the high end are capital-intensive facilities that include complex pre-processing systems and highly special-ized vessels that allow constant control of aeration, temperature, moisture and other factors. Lower technology facilities typically accept only yard debris while higher technology facilities may process materials such as sewage sludge and mixed municipal solid waste. All composting operation attempt to biologically decompose materials under controlled conditions to produce a stable humus product. Generally speaking, the more complex the feed stock, the more technology is required to control the conditions throughout the various processing stages. Typically, economics are the major determinant of the technology employed by a community in a composting operation. Below a certain level of technology. It is not possible to guarantee protection of public health and environmental quality (Golueke and Diaz, 1991).

Feedstocks

The composition of compost feedstocks is controlled by both collection and processing methods. Feedstocks suitable for composting include: the mixed organic fraction of MSW, leaves, brush, yard trimmings, biosolids (sewage sludge), forestry and other residuals, manure and agricultural residuals, food processing residue and other industrial compostable by-products. Source separation of MSW prior to composting is gaining popularity around the world (M.M. Dillon Ltd. and Cal Recovery Systems, 1990). Several European countries and many organizations in the United States are advocating source separation of organics and advocate prohibition of composting the mixed organic fraction of the solid wastestream due to concerns over heavy metals and other contaminants (Epstein *et al.*, 1992).

In yard waste composting, there are collection systems which minimize contaminants though use of specialized containers or bags, carefully designed schedulling, public education and other elements (Richard, 1991). It sewage sludge is included in the feedstock, it is critical that pretreatment standards are consistently adhered to (Richard, 1991). Due to the higher concentration of heavy metals in industrial sludge, another method of ensuring that product quality is not jeopardized is to restrict feedstocks to residential sludge (M.M. Dillon Ltd. and Cal Recovery Systems, 1990).

Pre-processing

Pre-processing of feedstock materials is required to remove undesirable materials and to prepare it for maximum decomposition and biological activity. Three major processes typically employed are particle size reduction, materials classification or separation, mixing and moisture conditioning (Beck, 1990). These activities are conducted in different sequences depending on the facility, and some steps may not be performed at all. Lower technology facilities handling yard waste may remove large contaminants manually and use a chipper to size-reduce materials. Higher technolgoy facilities typically employ a pre-processing line that includes hand sorting. Higher technology facilities typically employ a pre-processing line that includes hand sorting, magnetic separation, air classification, shredding, grinding and screening (R.W. Bech, 1990).

Processing

Certain aspects of product quality can be affected only by controlling or adjusting the compost process itself. For example, the population of pathogens in sewage sludge can be reduced to a safe level by maintaining the thermophilic temperature in compost mass over a sustained period (M.M. Dillon Ltd. and Cal Recovery Systems, 1990). The actual composting process is accomplished with a variety of available technologies including window with turning, static pile with forced aeration or in-vessel systems (Keener *et al.*, 1993). There have been many research efforts to study these various systems and determine operational data for the success of particular configurations.

Post-processing

Post-processing, includes curing, screening, magnetic separation, pneumatic separation and grinding, and is critical to assuring product quality and marketability. Through humification and mineralization, the curing process stabilizes the compost by reducing the presence of pathogens and allowing any uncomposted material to further decompose to a stable, mature product. Maturity is the extent to which the raw material has been stabilized during the composting process. It can be predicted through a number of means including chemical and physical analyses, microbiological assays, plant bioassays, spectroscopy and degree of huminification (Inbar *et al.*, 1990). Maturity is critical for the successful application in agriculture (Inbar, *et al.*, 1990); immature compost can cause severe damage to plant growth (Saviozzi *et al.*, 1988).

Benefits of Compost Application

Most benefits from compost application to the soil are related to increasing the organic matter content rather than as a fertilizer (Shiralipour *et al.*, 1992). Although there are concerns about potential public health hazards from the presence of pathogens and pollutants such as a trace metals and organic contaminants, compost has a positive effect on the physical and chemical property of soil as well as microbial populations such as rhizosphere organisms. It also suppresses population levels of nematodes that attack plants. Improvement in soil physical, chemical and biological characteristics by compost application will result in water and fertilizer, and herbicides/pesticides savings and will increase crop yields.

Composting in Developing Countries

Management of municipal waste is a major concern for governments and municipalities throughout the world. The heavy financial burden, the difficult logistical and social issues raised, and the environmental and health risks involved place waste management at the forefront of problems which must be dealt with by third world governments. In developing countries, cities often spend 30 to 50% of their operating budgets on solid waste management without even effectively providing adequate levels of service in terms of courage and affordability (Arosoroff and Bartone, 1987). Among the benefits of compost application in developing countries are creation of jobs and marketable products, reduction of environmental pollution, and potential for increased agricultural productivity through reuse of organic waste in rural areas surrounding municipalities (Arosoroff and Bartone, 1987).

Highly mechanized composting facilities common in Europe and the United States are not recommended for developing countries. Simple technologies such as windrow and static pile systems are usually adequate with a minimum of front-end processing because of higher organic and moisture content and smaller particles size of solid wastes in third world cities, and ambient temperature advantage in the tropics (Arosoroff and Bartone, 1987). Separation of glass metal and plastics by hand may be the only preparation required.

REFERENCES

Arlosoroff, S. and Bartone, C. (1987). Assisting Developing Nations. *BioCycle.* July: 43–45.

Beck, R.W. and Associates. (1990). Pierce Country report on alternative solid waste processing technologies. Submitted to the Pierce Country Utilities Department Tacoma, Washington.

Carr, J.S. and Cossu, R. (1990). *International Perspectives on Municipal Solid Wastes and Sanitary Landfilling*, Academic Press, San Diego.

Christopher, T. and Asher, M. (1994). *Compost This Book!* Sierra Club Books, San Francisco.

Epstein, E., Chaney, R.L., Henry, C. and Logan, T.J. (1992). Trace elements in municipal solid waste compost. *Biomass and Bioenergy*, **3(3–4):** 227–238.

Glenn, J. 1992. State of garbage in America–I. *BioCycle.* **33(4):** 46–55.

Goldstein, N. and Riggle, D. (1989). Healthy future of sludge Composting. *Biocycle.* **29(12):** 28–34.

Goldstein, N., Riggle, D. and Steuteville, R. (1992). Sludge composting. *BioCycle.* **30(12):** 28–34.

Goldstein, N., Riggle, D. and Steuteville, R. (1992). Sludge composting maintains growth. *BioCycle.* **33(12):** 49–56.

Golueke, C.G. and Diaz L.F. (1991). Low tech composting for small communities. In: *The BioCycle Guide to the Art and Science of Composting* (BioCycle Staff, Eds). J.G. Press, Inc. Emmaus, PA. Pp. 75–77.

Inbar, Y., Chen Y., Hadar, Y. and Hoitnik, H.A.J. (1990). New approaches to compost maturity. *BioCycle.* **31(12):** 64–69.

Kashmanian, R.M. (1993). Quantifying the amount of yard trimmings to be composted in the United States in 1996, *Compost Sec. Util.*, 1(3), 22.

Keener, H.M., Marugg, C., Hansen, R.C., Hoitink, H.A.J. (1993). Optimizing the efficiency of the composting process. In: *Science and Engineering of Composting.* (Eds: Harry A.J. Joitink and Harold M. Keener), Renaissance Publication, Worthington, Ohio. Pp. 59–94.

M.M. Dillion Ltd. and Cal Recovery Systems (1990). *Composting: A Literature Study.* Prepared for the Ontario Ministry of the Environment. Queen Printer for Ontario, Canada.

Richard, T. (1991). Composting methods and operations. In: *The BioCycle Guide to the Art and Science of Composting* (BioCycles Staff, Eds). J.G., Press, Inc., Emmaus, PA. Pp. 42–72.

Saviozzi, A., Levi-Minze, R. and Riffaldi, R. (1988). Maturity evaluation of organic waste. *BioCycle.* **29(3):** 54–56.

Shiralipour, A., McConnell, D. and Smith, W.H. (1992). Compost: Physical and chemical properties of soil as affected by MSW Compost application. *Biomass and Bioenergy,* **3(3–4):** 261–266.

Shiralipour, A., McConnell, D.B. and Smith, W.H. (1992). Compost: Physical and Chemical properties of soils as affected by MSW compost application. *Biomass and Bioenergy.* Vol. 3, Nos: 3–4, pp. 261–266.

U.S. Environment Protection Agency. (1990). Characterization of municipal solid waste in the United States: 1990 Update. Executive Summary. EPA/530–SW–90–042A. Washington, D.C. 15 pp.

U.S. Environmental Protection Agency (1994). Characterization of municipal solid waste in the United States: 1994 Update EPA-530–S–94–042.

11

Eco-friendly Technologies for Environmental Remediation/ Management

S. Kastury

INTRODUCTION

It is difficult to exactly pinpoint exactly when environmental conscience triggered environmental awareness and movements in the global community. Since the manufacture of synthetic chemicals began, the scientific and environmental community have expressed serious concerns about the disposal and improper management of hazardous toxic chemicals or wastes (U.S. National Research Council, 1994). In America environmental awareness has led to a movement resulting in hundreds, if not, thousands, of federal/state regulations over the last decades.

Over the past 15 years, some evidence has accumulated that ground water resource, which supplies more than 50% of drinking water in the US is threatened not only by excessive overdraws but also by past and present industrial, agricultural and commercial activities (O'Brien and Gene Engineers, 1988). In US, it was estimated that more than 300,000 sites may have contaminated soil or ground water requiring some form of remediation (EPA, 1993). Recent estimates of total cost of cleaning up these sites over the next 30 years may be as high as $ 1 trillion (O'Brien and Gene Engineers, 1988).

Until recently, coventional method of remediation of ground water at a given contaminated site were the traditional "pump and treat" systems. The "pump-and-treat " systems are the most common technical approach for ground water remediation in the United States. The "pump-and-treat system" involves the installation of several wells at the strategic locations at a contaminated site, pump the ground water, treat them with regulatory treatment standards (mostly to drinking water standards) and discharge the treated water back to the ground. Several recent studies have raised some questions about whether the so-called traditional "pump-and-treat" systems are capable of solving environment problems associated with ground water contamination. These studies indicate that conventional pump and treatment systems may not be able to remediate the ground water to required standards, removal of contaminants may not be economical and in some cases may be taking very long periods may be decades.

ECO-FRIENDLY TECHNOLOGIES

As a result of these concerns, a new trend has been developed looking for innovative alternate, relevant, appropriate and eco-friendly (IARAE) technologies for environmental remediation. Scientists, engineers, governmental agencies, industry and environmental groups representing affected citizens are looking for such eco-friendly technologies as a tool for effective environmental remediation. These groups have a concern that large amount of money is being wasted on ineffective remediation systems and that public health of current and future generations may be at risk due to these ineffective cleanups.

Environmental remediation is inherently, a very complex phenomenon. The contaminated sites have complex physico-chemical and geohydrological properties and complex behavior of contaminants in the subsurface have a significant impact on the type of IARAE technology. Some of these factors include:

1. Difficulties in characterizing the subsurface
2. Presence of light non-aqueous-phase liquids (LNAPLs) and dense NAPLs (DNAPLs)
3. Heterogeneity of subsurface environment
4. Public health concerns

IARAE technologies should be correctly chosen and implemented. IARAE technologies are new methods of remediation, sometimes these technologies are modifications or new applications of conventional methods. These technologies have the potential to significantly improve the efficiency of environmental remediation, especially when these technologies are suited for specific types of contaminants or when specific hydrogeologic environments are combined. While no technology can ensure the 100% achievement of cleanup standards, IARAE technologies, nevertheless, have the potential to increase the effectiveness and reduce the cost of environmental remediation.

While conventional "pump-and-treat" systems require a continuous energy input for pumping water and hence, energy-intensive, the IARAE systems may not require continuous energy inputs therefore, are energy efficient and cost effective, and hence highly suitable for tropical conditions. Tropical countries have an advantage over non-tropical countries, because of favorable environmental factors, countries like India should take advantage of these IARAE technologies for efficient and effective environmental remediation.

Some of the recommended IARAE technologies are outlined in this article. These technologies are currently being experimented and implemented at various contaminated sites in the United States. The technologies being implemented are:

1. Bioremediation
2. Phytoremediation
3. Bioventing
4. Composting
5. Bioreactors
6 Natural Attenuation (NA) of soils
7. Soil Vapor Extraction (SVE)

This article discusses important elements of some of these IARAE technologies.

Bioremediation

Bioremediation is a process where, naturally occurring micro-organisms (yeast, fungi and bacteria) are used to degrade or breakdown contaminants to less toxic or non-toxic substances. Micro-organisms, just like humans, eat and digest organic substances for nutrients and energy. Certain micro-organisms can digest organic substances that are hazardous to humans. (The type of micro-organisms present in the contaminated media is an important factor.) Because different micro-organisms degrade different types of compounds and survive under different conditions.

These micro-organisms can be broadly categorized as:

* Indigenous
* Exogenous

Indigenous micro-organisms are those micro-organisms which are native to the site. To stimulate growth of these indigenous micro-organisms, certain environmental factors (as outlined in Table 1) of the soil may need to be adjusted.

Table 1. Environmental Factors

• Oxygen	(>0.2 mg/l D.O <0.2 mg/l D.O for anaerobic)
• Nutrients	(Suggested C:N:P ratio of 120:10:1 {wt. Basis} O_2:C suggested ratio of 3:1)
• Temperature	15–45°C
• pH	6–9
• Moisture content	25–85% of moisture holding capacity

If micro-organisms that are needed to degrade the contaminant are not present in the soil, then micro-organisms from other locations, whose effectiveness has been tested in laboratories, are added to the contaminated soils (EPA, 1992). These are known as exogenous micro-organisms. Environmental factors (as outlined in Table 1) may need to be adjusted.

Table 2 outlines some important design criteria for bioremediation.

Table 2. Criteria for Bioremediation

- Environmental Factors
- Site Characterization
- Treatability Studies
- Biodegradation and Metabolism

Bioremediation is a combined process of respiration/biochemical reaction and enzyme producing microbes and end products. Two classes of biodegradation reactions takes place. These are:

* Aerobic

* Anaerobic

Aerobic systems involve usage of oxygen for biodegradation. Oxygen (O_2) is the terminal electron acceptor: At some contaminated sites, due to depletion of oxygen (O_2) and slow recharge of O_2 environmental conditions could become anaerobic Studies have shown that nitrate iron and manganese, sulfate and carbon dioxide (CO_2) can act as electron acceptors for the anaerobic conditions (EPA, 1996). The general degradation of hazardous compound by aerobic and anaerobic mechanisms can be unstarted by the following equations:

$$\text{Organics} + O_2 = \text{Biomass} + CO_2 + \text{Other inorganics}$$
$$NH_3 + CO_2 + H_2 + O_2 = \text{Biomass} + NO_3 + H_2O$$

** Some micro-organisms (e.g., chemoautotrophic aerobes or lithrophic aerobes) oxidize Inorganic compounds (NH_3, Fe or H_2S) to gain energy and fix CO_2 to build cell carbon.

Selected aerobic and anaerobic respiration involved in microbial metabolism organic matter is shown in Table 3 (EPA, 1991). Various anaerobic processes, as shown in Table 3 require a non-O_2 electron acceptor.

Table 3. Aerobic and Anaerobic Respiration in Microbial Metabolism

Process	Electron Acception	Metabolic Products
Aerobic	O_2	CO_2, H_2O
Anaerobic		
Denitrification	NO_3	CO_2, N_2
Iron Reduction	Fe	CO_2, Fe
Sulfate Reduction	SO_4	CO_2, H_2S

Bioventing

Bioventing is a process of injection of air into contaminated soils at low rates in such a way that soil O_2 concentrations are gradually increased and stimulate indigenous micro-organisms activity. Bioventing is most effective on organic contaminate, such as fuels and solvents. This is an in situ technology.

In addition to air, other nutrients can be pumped into contaminated soils in small amounts through injection wells. For example, nitrogen and phosphorous can be pumped for optimizing microbial growth. Micro-organisms will use the contaminants in the soil as food source and convert them to nonhazardous substances. The end products of this conversion reaction are CO_2 and water.

Studies have also shown successful results on nonfuel contaminants such as acetone, toluene, polycyclic aromatic hydrocarbons (PAHs)[8]. During bioventing, volatile organic compounds (VOCs) may escape from the soil before degradation occurs. This may be due to the fact that air from bioventing flows through the soils at a rapid rate. This can be prevented by bioventing in conjunction with air extraction, for this, air extraction wells are installed in conjunction with infection wells.

Phytoremediation

Phytoremediation is the use of higher plants to bioremediate contamination in soils, groundwater or sediments. This eco-friendly technology has been used in the past for wetlands to treat municipal sewage or neutralize acidic mine drainage. At present phytoremediation can be used for remediation of both organic and inorganic contaminants in soils and groundwater.

Unlike bioremediation which uses microbes, the phytoremediation process involves the usage of processes to change the form of the contaminants. The roots of plants take up contaminated water, nutrients and other compounds from the soil. Water moves through the plant system, to the leaves and eventually, to the atmosphere, due to the process of transpiration. Routine plant metabolic processes use water, nutrients and carbon dioxide in the presence of sunlight to synthesize organic compounds (which are moved through the plant system) for growth and storage of reserves. The micro-organisms thrive in the contact primarily, with the plant root systems and are supported by the products of the plants.

Table 4 indicates the suitable environmental characteristics for phytoremediation. Much of the biodegradation associated with phytoremediation takes place in a zone around the root system[9]. This zone is known as rhizosphere. The rhizosphere helps larger microbial populations than the surrounding soils. This enhanced microbial activity at the rhizo-sphere is responsible for degradation of certain contaminants.

Table 4. Phytoremediation

• In soil:	Fairly shallow, widespread low to medium concentration contaminants
• In groundwater:	Shallow (0–20′)
• Less expensive:	Tested for hydrocarbons, pesticides, VOCs
• Low installation and maintenance cost	
• High public community acceptance	
• Can clean chronic pollution sources (acid type)	
• Shorter time frame (2–3 yrs)	

Table 5 shows the removal of metals by various types of plant species. Plants can absorb or adsorb and accumulate contaminants either in their roots or in stems, leaves and fruits. This type of eco-friendly technology is best suited for remediation over a wide area. with fairly shallow contaminants in the low to medium concentrations. Poplar and willow trees can be planted as intercept barriers to protect surface water from agricultural runoffs or to remediate groundwater contamination.

Table 5. Examples of Metal Removal

Metal	Plant species
Zn	*Thlaspi calaminane*
	Viola species
Cu	*Aeolanthus bifonmifolius*
Ni	*Phyllantus serpentinus*
	Sebertia acuminata
Pb	*Bnassuca juncea*
CO	*Haumaniastrum robertii*

Bioreactors

Bioreactors may be defined as a containment systems/reactor used to create a three-phase (solid, liquid and gas) mixing condition to expedite the biodegradation of soil bound and water bound contaminants and biomass (usually bacteria). Different type of eco-friendly bioreactors include:

- Slurry bioreactors
- Fixed film bioreactors
- Suspended growth bioreactors

Fixed film bioreactors use either fixed, expanded or fluidized beds of inert media (plastic stone, sand, wood, ceramics or glass) or adsorptive media (granular activated carbon, resins). The contaminant is removed by biosorption and biodegradation. The suspended growth bioreactor is a standard technology useful for treating organic contaminants in aqueous and waste sludge systems. The reactors use microbial metabolism under aerobic, anaerobic or aerobic–anaerobic conditions to biosorp organic compounds and biodegrade them to residuals. The reactor configurations include sequential batch reactors (SBRs), completely mixed activated sludge systems, plug flow activated systems and aerobic/anaerobic digestors. These eco-friendly reactor systems require sufficient amounts of organic carbon in the stream or in the sludge to support a stable microbial culture in the bioreactor. A typical configuration could be about 5–10 lb influent BOD/day per 1,000 cubic feet or bioreactor volume and atleast a 100lb influent VSS/day per, 100 cubic feet of aerobic or high-rate anaerobic digester volume. Typical loading ranges for suspended growth reactors are given in Table 6[10].

Table 6. Loading Rates

Type of Reactor	Loading Rate
Aerobic sludge digester	100–300 lb, ss/day/1000 ft^3
Anaerobic sludge digester	40–200 lb, ss/day/1000 ft^3
Plug flow	20–40 lb,BOD/day/1000 ft^3
Completely mixed	50–120 lb, BOD/day/1000 ft^3
Extended aeration	10–25 lb, BOD/day/1000 ft^3

Table 7. Examples of Hazardous Constituents Oxidized by Micro-organisms

Micro-organism	Hazardous constituent (s)
Pseudomone putida	Phenol, TCE
P.cepacia	Phenol, TCE
Nitrosomonas europaea	Benzene, Ethylbenzene, Toluene
Cresols, Phenols	
Rhodococcus rhodochrous	TCE, Toluene, Ethylbenzene
Methylosinus trichosporium	TCE
Cunninghamella elegans	Toluene, Anthracene, Phenanthrene
Maphthalene	
Candida tropicalis	Phenols
Trichosporon cutaneum	Phenols, Cresols, Xylenes
Phanerochate chrysosporium	PCP, DDT, TNT

REFERENCES

U.S. National Research Council (1994). *Alternatives for Groundwater Cleanup*. National Academy Press, Washington, D.C., U.S.A (1994).

O'Brien and Gene Engineers Inc. (1988). Hazardous Waste Site Remediation: The Engineer's Perspective. Van Nostrand Reinbold, (1988).

EPA (1993). Cleaning up the Nation's Waste Files: Markets and Technolgoy Trends, EPA 542-R-92-012, Washington, D.C., EPA/OSWER. And O'Brien and Gene Engineers Inc. (1988).

EPA (1992). *A Citizen's Guide to Using Indigenous and Exogenous Microoganisms*. EPA/542/F-92/009, OSWER, March.

EPA Seminars (1996). Bioremediation of Hazardous Waste Sites Practical Approaches to Implementation, ORD/EPA/625/K-96/001, May. And Sims, R.C. (1990) Soil Remediation Techniques at Uncontrolled Hazardous Waste Sites, *J.Air Waste Management Assoc.*, 40(5): 703–722.

EPA, *Anaerobic Biotransformation of Contaminants in the Subsurface.* EPA/6000m-90/024, 1991(a).

McCauly, P.T.; R.C. Brenner, F.V. Kremer, B.C. Alleman, and D.C. Beckwith (1994). Bioventing Soils Contaminated with Wood Preservatives. In: Symposium on *Bioremediation of Hazardous Water R&D and Field Applications,* EPA/600/R-94/075.

EPA Seminars, (1996). Bioremediation of Hazardous Waste Sites, *Practical Approaches to Implenentation* ORD/EPA, EPA/625/K-96/001, (May).

O'Brien and Gene Engineers. *Innovative Engineering Technologies for Hazardous Waste Remediation.* Van Nostrand Reinhold. (1994).

12

Production of Ethanol from Lignocellulosic Materials Using *Clostridium thermocellum*— A Critical Review

G. Seenayya, Gopal Reddy, M. Sai Ram, M.V. Swamy and K. Sudha Rani

INTRODUCTION

Energy is obtained from biomass either by direct combustion or by microbial action. Amongst microbial conversion of biomass to fuels, production of ethanol has a great potential and is one of the best eco-friendly technologies for biomass conversion into energy. Energy ranks alongside population growth and food supply as central obstacles to economic growth and social welfare.

The technological and industrial boom before 1973 in developed countries is mainly due to the availability of cheap energy, i.e. oil. The steep escalation of oil prices in 1973 and 1979, and subsequent price fluctuations have created a concern and awareness of the finite nature of these fossil fuels leading to a search for alternative sources of energy. In this context, large scale ethanol production from biomass is considered as an alternative either as a fuel, a fuel additive, or as a chemical feed stock. This also alleviates the environmental pollution problems.

PRESENT FERMENTATION ETHANOL PRODUCTION

More thar 95% of fermentation ethanol produced world wide employs yeast, *Saccharomyces cerevisiae* and its related species (Rosillo-Calle and Hall, 1987; Slapack et al., 1985). Ethanol yields from selected strains of *S. cerevisiae* using molasses as feed material range between 85%–90% of the theoretical yield (Basappa, 1987).

Theoretically

$C_8H_{12}O_6$ ⟶	$2C_2H_5OH$	+ $2CO_2$
Glucose	Ethanol	Carbon dioxide
1 mole	2 moles	2 moles
1 g	0.51 g	0.49 g

In Practice

$$1\text{g glucose} \xrightarrow{\underline{S.cerevisiae}} 0.43\text{–}0.46 \text{ g ethanol}$$

However, yeasts have a narrow substrate spectrum and can only ferment glucose, maltose, sucrose and fructose. They cannot ferment xylose, starches and cellulosics (Lovitt et al., 1988).

A number of processes employing a mesophilic bacterium. *Zymomonas mobilis* for ethanol production have been developed. Although *Z. Mobilis* has higher ethanol tolerance and production abilities, *S. cerevisiae* is preferred, as the *Z. mobilis* requires maintenance of neutral pH and sterilization of media, etc. (Swings and Deley, 1977).

Among the top three ethanol producing countries in the world, India and Brazil use molasses and USA uses maize starch after enzymatic hydrolysis as feedstock materials for ethanol production. However, sugar and starch materials are in short supply and are required for many alternative uses. Therefore, the major breakthrough in large scale production of ethanol will depend upon the use of less expensive and renewable feedstock such as lignocellulosics (Lipinsky, 1981, Lovitt et al., 1988).

Biomass for Ethanol Production

The lignocellulosic feed stocks have, 30%–50% cellulose, 20%–30% hemicellulose and 20%–25% lignin. Cellulose is linear hydrophilic polymer made up of glucose sub units linked by β-1,4, glycosidic bonds. The hermicelluloses are heteropolymers of pentoses (xylose, mannose, arabinose), hexoses and glucuronic acid. Lignin is made up of phenyl propane units, methoxylated and linked in various ways. A major physico-chemical feature limiting the direct fermentation of biomass is its polymeric nature celluloses and hemicelluloses are fermented by certain anaerobic bacteria, whereas lignin is totally recalcitrant and limits the activity of microbial celluloses and hemicelluloses by stearic hindrance. However, alkali treatment of lignocellulosic materials removes lignin and other inhibitory materials, and renders the biopolymers accessable to enzymatic degradation (Vallander and Eriksson, 1990; Lynd et al., 1991) and ethanol production.

It is estimated that global synthetic rate of cellulose is approximately 4 × 10 tones per year (Singh and Hayashi, 1995). A comperhensive data on the potential availability of the biomass resources in India, their ease of collection alternative uses and applications are not available. As per the survey of Technology Information Forecasting and Assessment Council (TIFAC, 1991) of the Department of Science and Technology, about 380 million Mt of biomass from major 12 crops is generated in India during 1989–90 and the paddy residues constitutes more than 50% of the total biomass.

CONVERSION OF CELLULOSIC BIOMASS TO ETHANOL

The conversion of cellulosic biomass to ethanol is mainly carried out either by the conventional two step approach or by a novel single step process (Fig 1). The conventional two step approach includes acid or enzymatic hydrolysis of the substrate followed by yeast fermentation. Acid hydrolysis

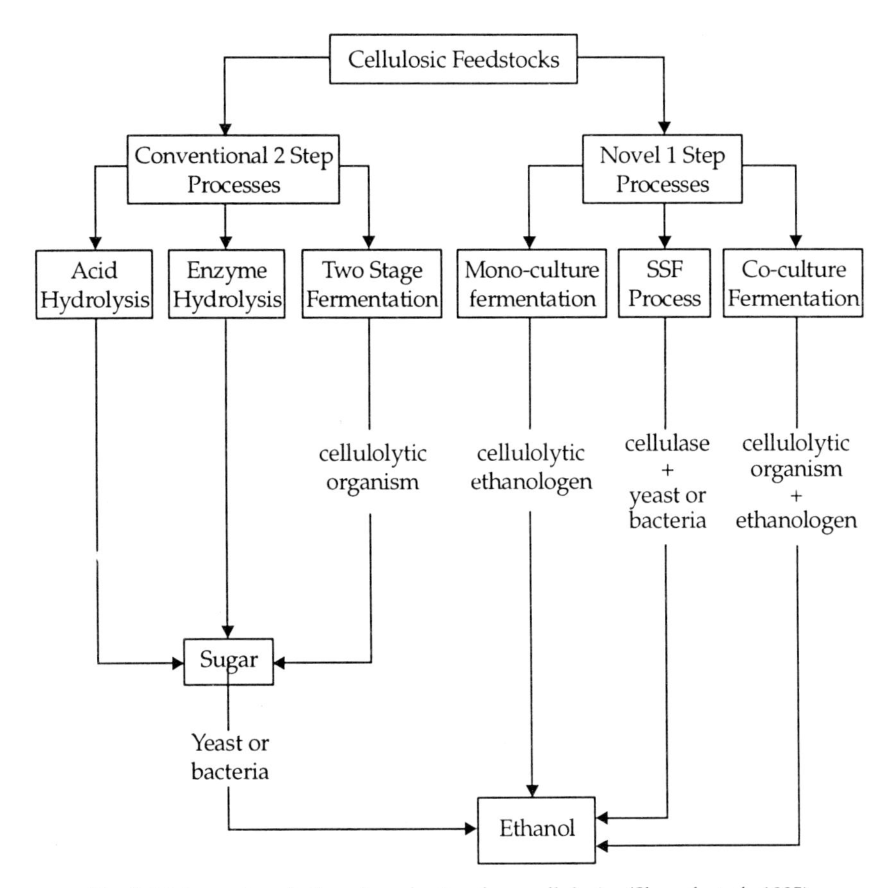

Fig. 1. Major routes of ethanol production from cellulosics (Slapack *et al.*, 1985).

is hindered mainly by glucose yields and corrosion of the equipment. Enzymatic hydrolysis, which usually employs cellulases from *Trichoderma reesel*, achieves higher substrate conversion yields, but its production is very expensive (Vallander and Eriksson, 1990). The single step processes either by monoculture or by coculture fermentations are highly economical since cellulase production is made in situ and eliminates the need for separate fermentor. For the economic conversion of cellulosic biomass to ethanol. The hemicellulose fraction must also be effectively fermented. Clostridium thermocellum is the only bacterium shown to date to produce true cellulase as well as xylanase system (Duong et al., 1983; Gilbert and Haziewood, 1993) and is capable of producing ethanol as a major fermentation product.

CLOSTRIDIUM THERMOCELLUM

Clostridium thermocellum is an anaerobic, thermophilic, celluloytic and ethanologenic bacterium. Several strains of *C. thermocellum* have been studied for their ability to produce ethanol directly from cellulosic biomass (Avgerinos et al., 1981; Duong *et al.*, 1983; Bender *et al.*, 1985; Freier *et al.*, 1988; Lynd, 1989; Mori 1990a; Sai Ram et al., 1991a, 1991b; Sudha Rani et al., 1996a. All the strains of *C. thermocellum* ferment cellulose and cellobiose and produce ethanol, acetate, lactate, CO_2 and H_2 as the end products. Few strains of *C. thermocellum* have the ability to utilize various mono, di and polysaccharides and sugar alcohols. The production of cellulose by *C. thermocellum* is one of the most important attributes for ethanol production from cellulose. The cellulose system of *C. thermocellum* is similar to *Trichoderma reesel* in being multicomponent, extracellular and hydrotyzes carboxymethycellulose, microcrystalline cellulose and hemicellulose (Coughian and Ljungdahi, 1988; Beguin and Aubert, 1994) and has higher specific activity on crystalline cellulose than fungal cellulose (Lynd, 1989).

Due to the unique nature of *C. thermocellum* cellulose, cloning of cellulose genes from *C. thermocellum* into *Z. mobilis, S. cerevisiae* and *Escherichia coli* (Singh and Hayashi, 1995) has been reported. E.Coli has been the most common host for cloning of different cellulose genes. Most of the work on cloning has been carried out mainly by the group of scientists at Pasteur Institute in France (Beguin and Aubert, 1994). They have demonstrated that the cellulose genes are not clustered and constitute about one-third of the *C. thermocellum* genome. However, the cellulolytic system with the unique ability to degrade crystalline cellulose, has not been reconstructed, despite the cloning of about 20 genes (Lowe *et al.*, 1993).

Cellulose degradation by *C. thermocellum* is accompanied by the production of water insoluble "Yellow Affinity Substance" (YAS). In the cellulolytic activity of *C. thermocellum*, cellobiose and cellodextrins are produced which enter into the metabolism of the organism for the production of ethanol. (Duong *et al.*, 1983; Nochur *et al.*, 1992; Strobe *et al.*, 1995). The wild strains of *C. thermocellum* reported so far produced 0.08–0.29g ethanol per gram glucose equivalents fermented (Duong *et al.*, 1983; Slapack *et al.*, 1985; Freier *et al.*, 1988; Mori, 1990a; Sai Ram *et al.*, 1991b) and are tolerant upto 1.5% (v/v) ethanol (Herrero and Gomez, 1980; Tailliez *et al.*, 1989a, 1989b; Sai Ram and Seenayya, 1991).

Though several important distinguishing features of *C.thermocellum* for ethanol production from cellulosic biomass have been identified, certain limitations exits which prevent the utilization of these strains for industrial scale production of ethanol from renewable resources (Slapack *et al.*, 1985; Rogers, 1986; Lovitt *et al.*, 1988; Lynd, 1989; Lowe *et al.*, 1993; Beguin and Aubert, 1994). Cellobiose and glucose, formed during the cellulolytic activity limit the cellulose synthesis and becomes a rate limiting step. This has been overcome by coupling cellulose fermentation by *C. thermocellum* with a hexose and pentose fermenting secondary bacterium i.e. coculture fermentation (Avgerinos *et al.*, 1981; Wang *et al.*, 1983; Zeikus *et al.*, 1983; Ljungdahl *et al.*, 1981; Carreira and Ljungdahl, 1983; Mori, 1990a). The fraction of the substrate utilized for the production of acetate and lactate lowers the substrate conversion to ethanol. Another major limitation in the utilization of *C. thermocellum* strains for ethanol production is their marked intolerance to ethanol.

RECENT DEVELOPMENTS

Recently new strains of *C. thermocellum* with desirable properties were developed or isolated. Sato et al., (1992) have isolated *C. thermocellum* 1–1–B which produced 86.8 mM (0.39 g/g) ethanol at a substrate concentration of 1% in the presence of 1.4% yeast extract. Enhancement in ethanol production by *C. thermocellum* strain SS8 from 0.25 to 0.39 g/g of cellulosic substrate in the presence of 0.15 mM sodium azide or 7% polyethylene glycol (metabolic inhibitors) along with significant repression in acetic acid formation was observed by Sudha Rani *et al.*, (1994). *C. thermocellum* strains SS19, SS21 and SS22 producing 0.32g, 0.37g and 0.33 g ethanol per gram substrate consumed with ethanol to acetate (E/A) ratio of 1.72, 2.21 and 2.45, respectively have been isolated in our laboratory. These yields are considered to be high among the wild strains of *C. thermocellum* reported so far (Sudha Rani *et al.*, 1996a).

Improved ethanol tolerance upto 3.2% (v/v) for *C. thermocellum* strain C9 (Herrero and Gomez, 1980), 4.8% and 5.0% (v/v) for strain LD1 and 657 (Tailliez *et al.*, 1989a, 1989b), 5.0% (v/v) for strain GUV_2 and TN_{45} (Sudha Rani *et al.*, 1996b) and 6% (v/v) for strain S-7 (Wang *et al.*, 1983) have been reported. The strains improved for high ethanol tolerance had increased ethanol production and E/A ratio. The ethanol tolerance of the newly isolated strains SS21 and SS22 (Sudha Rani *et al.*, 1996a) in our laboratory was also very high (4% and 5%v/v) when compared with the other wild strains of C. *thermocellum* and almost equal to the improved strains reported so far.

Mori (1990a, 1990b) has isolated a high cellulose producing C. *thermocellum* strain YM4 and improved for further hyper cellulase productivity to attain more effective cellulose decomposition. The mutant strain Y-25a degrade 10kg/1 of avicel within 20hr.

The ability of C. *thermocellum* strain to convert the cellulosic (6C sugars) fraction of biomass to ethanol is only one-half to two-third of the biomass conversion effciency. This is because most natural biomasses such as paddy straw, corn stover etc. contain about one-third to one-half hemicellulosic (5C sugars) materials. If the hemicellulose cannot be utilized effectively, the overall conversion efficiency would be quite unfavourable with respect to process economics. Recently, broad saccharolytic ability strains such as strain JW20 (Freier *et al.*, 1988), strain YM4 (Mori, 1990a) have been reported. We have shown in our laboratory that the strain SS8 (Sai Ram and Seenayya, 1989) and strain SS22 (Sudha Rani *et al.*, 1996a) are effective in hydrolysis of hemicelluloses to ethanol in addition to cellulose and other carbohydrates.

All the strains of C. *thermocellum* so far reported have shown poor growth on crude-biopolymers and delignification with alkali treatment has increased the biodegradation of the substrate and ethanol yields (Ng *et al.*, 1981, Kundu *et al.*, 1983; Saddler and Chan, 1984; Sai Ram and Seenayya, 1991). The lignaceous components of the biomass are not being hydrolyzed and fermented by the strains of C. *thermocellum*. On the contrary, these substances inhibited the hydrolysis and fermentation process, since the fermentation efficiency of delignified material was high (Wang *et al.*, 1983; Kundu *et al.*, 1983; Sai Ram and Seenayya, 1991).

In the conversion of biomass to ethanol by C. *thermocellum*, most of the studies were conducted using pure cellulosic materials such as filter paper, avicel solkafloc, tissue paper etc., at substrate concentrations around 1.0%. When experiments were conducted by employing high ethanol yielding strains at high substrate concentrations, the ethanol yields decreased with increase in acetate production and accumulation of reducing sugars (Wang *et al.*, 1983; Tailliez *et al.*, 1991a, 1991b; Sai Ram and Seenayya, 1991).

To utilize accumulated reducing sugars, a second anaerobic and thermophilic bacterium (*C. thermosaccharolticum or C. thermohydrosulfuricum or Thermoanaerobacter ethanolicus*) is utilized. When high yielding ethanol tolerant mutant strain C. *thermocellum* S-7 and reducing sugar utilizing *C.thermosaccharolyticum* HG-4 were used in fed batch fermentation, 85% of the theoretical ethanol yield was obtained from 60 g/l solkafloc utilised out of 80 g/l taken (Wang *et al.*, 1983). However, when these strains were employed in the fed-batch fermentation with 100 g/l of non chemically treated corn stover, the substrate degradation was 37% the ethanol yield has decreased to 50% of the theoretical yield and acetate production has increased. The decrease may possibly be due to the toxicity of the lignaceous components (Avgerinos *et al.*, 1981).

Tailliez *et al.* (1989a 1989b) observed 14.5g/l and 12.7 g/l ethanol from 63 g/l and 48.3 g/l substrate (Whatman no. 1 filter paper) consumed, by ethanol tolerant mutant strains LD1 and 647, respectively. This accounts to 45% and 52% of the theoretical ethanol yield.

In our laboratory, a moderately ethanol tolerant (1.5% v/v) and abroad saccharolytic *C. thermocellum* strain SS8 produced 8.6 g/l of ethanol from 37.5 g/l of delignified bamboo pulp degraded out of 75 g/l taken, showing 47% of theoretical ethanol yield (Sai Ram and Seenyya, 1991). Two high ethanol tolerant strains SS21 and SS22 (4% and 5% v/v) produced 21.47 g/l and 23.57g/l ethanol from 82.6g/l and 87.3 g/l alkali extracted paddy straw degraded out of 100 g/l taken, showing 51% and 53% of theoretical ethanol yield (Sudha Rani, 1996).

In evaluating *C. thermocellum* strains for ethanol production, it has been assumed that ethanol tolerance comparable to yeast is required. The requirement for high ethanol tolerance is likely to apply if separation technologies which are designed for the ethanol concentrations tolerated by yeasts are employed. Recently, Leeper (1992) has compared the energy requirements of several ethanol recovery processes and showed that by integration of membrane technology separations into product recovery process, it is possible to recover ethanol at 4% (w/v) or 5% (v/v) with the same energy input as that of conventional distillation processes at 9% (w/v) or 13% (v/v) (Leeper, 1992).

CONCLUDING REMARKS

Large scale production of fermentation ethanol from sugar or starch substrates has limited potential. The non-availability of these substrates in abundance is the main limitation and this can be overcome by utilizing the abundant, renewable and less expensive lignocellulosic biomass after pretreatment. *C. thermocellum* strains are capable of fermenting the delignified biomass directly to ethanol. The limitations such as low ethanol tolerance, low ethanol yields, accumulation of reducing sugars, etc., associated in employing *C.thermocellum* strains have been overcome partially in the laboratory scale experiments.

The improved strains of *C.thermocellum* can be utilized in monoculture or coculture fermentation for economic production of ethanol from biomass to provide a partial substitution for fossil fuels. However, pilot scale experiments have to be conducted with integration of membrane separation technologies to assess the problems involved in the direct conversion of cellulosic biomass to ethanol.

REFERENCES

Anonymous, (1991). Technology, Inoformation, Forecasting & Assessment Council. Biomass generation and utilization. Department of Science and Technology, New Delhi. p.1–185.

Avgerinos, G.C., Fang, H.Y., Biocic, I. and Wang, D.I.C. (1981). A novel single step microbial conversion of cellulosic biomass to ethanol. p.119–124. In *Advances in Biotechnology,* Vol.2, M. Moo-young and C.W. Robinson (eds). Pergamon Press, Toronto.

Basappa, S.C. (1987). *Zymomonas mobilis* as an alternate organism for ethanol production. National Symp. Proceed. In: Yeasts for alcohol and fat production. p.32–51.

Beguin, P. and Aubert, J.P. (1994). The biological degradation of cellulose. *FEMS Microbiol. Rev.* 13: 25–58.

Bender, J., Vatcharapijarn, Y. and Jeffries, T.W. (1985). Characteristics and adaptability of some new isolates of *Glostridium thermocellum. Appl Environ. Microbiol.* 49 : 475–477.

Carreira, L.H. and Ljungdahl, L.G. (1983). Production of ethanol from biomass using anaerobic thermophilic bacteria. p. 1–30. In: D.L. Wise (ed.), *Liquid Fuel Developments.* CRC series in Bioenergy Systems. CRC Press Inc., New York.

Coughlan, M.P. and Ljungdahl, L.G. (1988). Comparative biochemistry of fungal and bacterial cellulolytic enzyme systems, p.11–30. In Proceedings of FEMS Symposium No. 43. *Biochemistry and Genetics of Cellulose Degradation.* J.P. Aubert, P.Beguin and J. Miller (ed.), Academic press, London and New York.

Duong, T.V., Johnson, E.A. and Demain, A.L. (1983). Thermophilic, anaerobic and cellulolytic bacteria. Top. Enzyme. Ferment. *Biotechnol.* 7: 156–195.

Freier, D., Mothershed, C.P. and Wiegel, J. (1988). Characterization of *Clostridium thermocellum* JW20. *Appl. Environ. Microbiol.* 54: 204–211.

Gilbert, H.J. and Hazlewood, G.P. (1993). Bacterial cellulases and xylanases, *J.Gen Microbiol.* 139: 187–194.

Herrero, A.A. and Gomez, R.F. (1980). Development of ethanol tolerance in *Clostridium thermocellum:* Effect of growth temperature. *Appl. Environ Microbiol.* 40: 571–577.

Kundu, S., Ghosh, T.K. and Mukhopadhyay, S.N. (1983). Bioconversion of cellulose into ethanol by *Clostridium thermocellum*-product inhibition *Bioeng.* 25: 1109–1126.

Leeper, S.A. (1992). Membrane separations in the recovery of biofuels and biochemicals: An update review. p. 99–194. In *Separation and Purification Technology.* M. Dekker (ed.), N.N.L. and J. M. Calo. New York.

Lipinsky, E.S. (1981). Chemicals from biomass: Petrochemical substitution options. *Science.* 211: 1465–1471.

Ljungdahl, L.G., Carreira, and Weigel, J. (1981). Production of ethanol from carbohydrates using anaerobic thermophilic becteria. Ekman-Daus. *Int. Symp Wood Pulping Chem* 4: 23–28.

Lovitt, R.W., Kim, B.H., Shen, G.J. and Zeikus, J.G. 1988. Solvent, production by Microorganisms. CRC Crit. Rev. Biotechnol. 7: 107–186.

Lowe, S.E., Jain, M.K., Zeikus, J.G. 1993. Biology, ecology and biotechnological applications of anaerobic bacteria adapted to environmental stresses in temperature, pH, salinity or substrates. Microbial. Rev. 57: 451–509.

Lynd,. L.R. 1989. Production of ethanol from lignocellulosic materials using thermophilic bacteria. Critical evaluation of potential and review. Adv. Biochem Engg/Biotechnol. 38: 1–52.

Lynd, L.R., Cushman, J.H., Nichols, R.J. and Wyman, C.E. 1991. Fuel ethanol from cellulosic biomass Science. 251: 1318–1323.

Mori, Y. 1990a. Characterization of a symbiotic coculture of *Clostridium thermohydrosulfuricum* YM3 and *Clostridium thermocellum* YM4. Appl. Environ. Microbiol 56: 37–42.

Mori, Y. 1990b. Isolation of mutants of *Clostridium thermocellum* with enhanced cellulase production. Agric. Biol. Chem. 54: 825–826.

Ng, T.K., Ben-Bassat, A. and Zeikus, J.G. 1981. Ethanol production by thermophilic bacteria. Fermentation of cellulosic substrates by cocultures of *Clostridium thermocellum* and *Clostridium thermohydrosulfuricum*. Appl. Environ. Microbial. 41: 1337–1343.

Nochur, S.V., Jacobson, G.R., Roberts, M.F. and Demain, A.L. 1992. Mode of sugar phosphorylation in *Clostridium thermocellum*. Appl. Biochem. Biotechnol. 33: 33–41.

Rogers, P. 1986. Genetics and biochemistry of *Clostridium* relevent to development of fermentation processes. p. 1–60. In: A.I. Laskin (ed.), In Advances in Applied Microbiology, Vol. 31, Academic Press, Inc., New York.

Rosillo-Calle, F. and Hall, D.O. 1987. Brazilian alcohol: Food versus fuel. Biomass 12: 97–128.

Saddler, J.N. and Chan, M.K.H. 1984. Conversion of pretreated lignocellulosic substrates to ethanol by *Clostridium thermocellum* in mono-and co-culture with *Clostridium thermosaccharolyticum* and *Clostridium thermohydrosulfuricum*. Can. J. Microbial. 30: 212–224.

Sai Ram, M. and Seenayya, G. 1989. Ethanol production by *clostridium thermocellum* SS8, a newly isolated thermophilic bacterium. Biotechnol. Letts. 11:589–592.

Sai Ram, M. and Seenayya, G. 1991. Production of ethanol from straw and bamboo pulp by primary isolates of *Clostridium thermocellum*. World J. Microbiol. Biotechnol. 7: 372–378.

Sai Ram, M., Rao, C.V. and Seenayya, G. 1991a. Characteristics of *Clostridium thermocellum* strain SS8, a broad saccharolytic thermophile. World J. Microbiol. Biotechnol. 7: 272–275.

Sai Ram, M., Swamy, M.V. and Seenayya, G. 1991b. Fermentation characteristics of ethanol producing isolates of *Clostridium thermocellum*. Indian J. Microbiol. 31: 175–180.

Sato, K., Goto, S., Yonemura, S., Sekine, K., Okuma, K., Takagi, T., Hon-nami, K. and Saiki, T. 1992. Effect of yeast extract and vitamin B12 on ethanol production from cellulose by *Clostridium thermocellum* 1-1-B. Appl. Environ. Microbiol. 58: 734–736.

Singh, A. and Hayashi, K. 1995. Microbial cellulases: Protein architecture, molecular properties and biosynthesis. Adv. Microbiol. 40: 1–44.

Slapack, G.E., Russell, I. and Stewart, G.G. 1985. Thermophilic bacteria and thermotolerant yeasts for ethanol production, NRCC No. 24410. Project Report Submitted to Division of Energy, NRC, Ottawa, p.1–404.

Strobe, H.J., Caldwell, F.C. and Dawson, K.A. 1995. Carbohydrate transport by the anaerobic thermophile *Clostridium thermocellum* LQRI. Appl. Environ. Microbiol. 61: 4012–4015.

Sudha Rani, K. 1996. Single step conversion of biomass toethanol - Substrate conversion efficiencies and ethanol tolerance of the natural isolates of *clostridium thermocellum* strain SS21 and SS22, Ph.D. Thesis, Osmaina University, India (submitted).

Sudha Rani, K. Swamy, M.V. and Seenayya, G. 1996a. High ethanol production by new isolates of *Clostridium thermocellum*. Biotechnol. Lett. 18: 957–962.

Sudha Rani, K., Swamy, M.V., Sunitha, D., Haritha, D. and Seenayya, G. 1996b. Improved ethanol tolerance and production in the strains of *Clostridium thermocellum*. World. J. Microbiol. Biotechnol. 12: 57–60.

Sudha Rani, K., Swamy, M.V., Sunitha, D. and Seenayya, G. 1994. Enhancement of ethanol production in the presence of metabolic inhibitors in *Clostridium thermocellum* strain SS8. Biotechnol. Letts. 16: 201–204.

Swings, J. and Deley, J. 1977. The biology of Zymomonas. Bacteriol. Rev. 41: 1–46.

Tailliez, P., Girard, H., Longin, R., Beguin, P. and Millet, J. 1989a. Cellulose fermentation by an asporogenous mutant and an ethanol tolerant mutant of *Clostridium thermocellum*. Appl. Environ. Micorbial. 55: 203–206.

Tailliez, P., Girard, H., Millet, J. and Beguin, P. 1989b. Enhanced cellulose fermentation by an asporogenous and ethanol tolerant mutant of *Clostridium thermocellum*. Appl. Environ. Microbial. 55: 207–211.

Vallander, L. and Eriksson, K.E.L. 1990. Production of ethanol from lignocellulosic materials: State of the art Adv. Biochem. Engg/Biothechnol. 2: 69–95.

Wang, D.I.C., Avgernios, G.C., Biocic, I., Wang, S.D. and Fang, H.Y. 1983. Ethanol from cellulosic biomass. Philos. Trans. R. Soc. Lond. B. Biol. Sci 300: 323–333.

Zeikus, J.G., Ng, T.K., Ben-Bassat, A. and Lamed, R.J. 1983. Use of co-culture in the production of ethanol by the fermentation of biomass. U.S. patent. 4, 400:470.

13

Cellulose Conversion to Ethanol by a Mesophilic Cellulolytic Bacterium

T. Vijaya and D.V. Dev

INTRODUCTION

Cellulolytic mesophilic bacteria ferment cellulosic biomass to ethanol at mesophilic temperatures. The bacteria hydrolyze both cellulose and hemi-cellulose substrate to their oligodextrins and monomeric components. Cellulolytic *Clostridium* species are able to ferment cellulose and cellobiose producing ethanol as the major fermentation product (Lynd, 1989, Lynd *et al.*, 1991) and except for a few species the difference seems to be in their ability to assimilate glucose and xylose (Giuliano and Khan *et al.*, 1984, Khan *et al.*, 1987). The major problem in the conversion of biomass to ethanol by *Clostridium* species is the accumulation of xylose, cellobiose and glucose which repress cellose fermentation resulting in low ethanol yields (Gardon *et al.*, 1978) and their marked intolerance to ethanol (Herrero and Gomez, 1980). These problems might be alleviated by co-culture fermentations and isolating cultures which are more efficient (Avgerinose *et al.*, 1981). Therefore, it is helpful to identify species which are more efficient than existing in the bioconversion of cellulose or more adaptable to practical fermentation conditions. In our investigation *(Clostridium tertium)* has been isolated which can ferment cellulosic biomass to ethanol (Vijaya and Dev 1997, 1999).

This article reports the studies made on substrate utilization, ethanol productivity, effect of pH, soluble sugars on the fermentation of cellulose by *Clostridium tertium* and also the nature of enzymes produced.

MATERIALS AND METHODS

Clostridium tertium used in this study has been isolated from municipal sewage mud on the medium containing (in g/l) $K_2HPO_4.3H_2O$, 1.0; NH_4Cl, 1.0; KCl, 0.5; $MgSO_4.7H_2O$, 0.5; L-cysteine hydrochloride monohydrate, 0.15; trypticase, 0.5, yeast extract (Difco) 0.5; Whatman No. 1 filter paper 4 and 20 ml trace metal solution (Ferguson and Mah, 1983); resazurin, 0.001g and 15 g of agar. The pH was adjusted to 7.2. This culture was maintained on the same medium without cellulose.

All the tests were conducted in serum vials of 120 ml capacity containing 20 ml of the prereduced medium in N_2 atmosphere. The media were sterilized by autoclaving at 121°C for 1 h. Sugar solutions were filter-sterilized before being added to the sterile medium. A three day old culture grown on 0.4 % cellulose was used as inoculum at 5 % v/v concentration. All the incubations were carried out at 34 °C ±2 without shaking. pH was maintained at 7.2 with sterile anaerobic 3N NaOH. The undegraded cellulose was measured according to the method of Weimer and Zeikus (1977). Reducing sugars were determined by DNS method described by Miller (1959). Ethanol and acetic acid were determined using chromosorb 101 column on a Aimil 5700 Nucon gas chromatograph equipped with a flame ionization detector.

For enzyme assay starch or cellulose (4.0 g/l) was used as substrate. After fermentation, the broth was centrifuged at 10,000x g for 30 min at 4 °C and the supernatant after dialysis was used as the enzyme source. CMCase/amylase activity was determined by incubating 1 ml of enzyme solution with 1 ml of 1% CMC (carboxy methyl cellulose, sigma) or soluble starch in 3 ml 0.1 M acetate buffer, pH 6.0 for 20 min at 34°C. One unit of enzyme is expressed as 1 µmol of reducing sugar equivalents released per hour.

RESULTS AND DISCUSSION

Clostridium tertium fermented efficiently (>80 % substrate at 10 g/l) a variety of crystalline, pure and agricultural cellulosic materials such as Avicel, filter paper, native cotton, paddy straw, sugarcane bagasse, bamboo pulp, tissue paper with light yellow pigment production. The products of fermentation include ethanol (0.23 to 0.27 g/g of substrate consumed), acetic acid, CO_2, H_2 and lactic acid. Ethanol to substrate ratio (E/S) increased to 0.29 when nitrogen was replaced by hydrogen in serum vials. Growth on soluble and substituted cellulose such as carboxy methyl cellulose was slow, producing acetic acid as the major fermentation product with an ethanol to substrate ration 0.08.

Clostridium tertium had a broad saccharolytic ability and grew on cellulose, cellobiose, glucose, xylose, lactose, arabinose, maltose, sucrose, galactose, mannitol, mannose, fructose, inulin and starch, producing 0.14 g to 0.26 g ethanol/g substrate consumed.

At the concentration of glucose or cellobiose tested, cellulose fermentation by *Clostridium tertium* was not affected during the first 48 hours. After 48 hours as the concentration of glucose increased, the cellulose fermentation was inhibited linearly resulting in 36 % of cellulose utilization at 0.5 % glucose. On the other hand, at 0.05 % cellobiose concentration, cellulose fermentation was about 70 % and further increase in the concentration of cellobiose had little effect (Table 1).

Table 1. Effect of different concentrations of glucose and cellobiose on % cellulose fermented (percentage degradation after 5 days)

Sugar added	Concentration (%)						
	0	0.05	0.1	0.2	0.3	0.4	0.5
Glucose	93	85	86	68	62	45	34
Cellobiose	95	67	65	66	68	67	61

At low substrate concentration, i.e., 10 g/l, pH had little effect and at higher concentrations of substrate, i.e., 30 g/l, the amount of cellulose fermented, reducing sugars accumulated and ethanol formed, varied depending on the pH of the medium (Table 2). When pH was not maintained, the fermentation products lowered the pH to 5.7 leading to the cessation of growth. Maintaining the pH at 7.2 by periodic injection of sterile 3N NaOH for every 24 hr resulted in the increase of biomass cellulose fermentation, utilization of accumulated sugars, production of ethanol and acetic acid. The organism exhibited high NaCl tolerance, OD_{520} increased up to 20 g/l NaCl and a further increase in NaCl concentration resulted in decreased ethanol yield.

Table 2. Effect of pH on the fermentation of cellulose by *Clostridium tertium* at different concentrations (incubation period 10 days)

Sl. No	Cellulose g/l	Ethanol g/l	Aceticacid g/l	Reducing sugars g/l	Cellulose fermented g/l	E/S ratio
1.	a 10.0	2.40	2.06	0.2	9.5	0.25
	b 10.0	2.41	1.88	0.1	9.6	0.25
2.	a 30.0	4.72	2.66	9.3	27.48	0.24
	b 30.0	7.48	3.11	3.3	29.28	0.27
3.	a 50.0	3.47	2.68	6.20	23.98	0.18
	b 50.0	7.66	3.38	4.48	37.58	0.22

a=pH not maintained; b=pH maintained at 7.2

Clostridium tertium produced both CMCase and Avicelase when grown on cellulose and the enzymes were repressed when grown on cellobiose or glucose. About 80 % of the cellulase activity was extracellular. Unlike CMCase, Avicelase activity was increased about two times in the presence of 5 mM DTT and 10 mM Ca^{2+}. The organism produced both cellulase and amylase on starch when cellulose adapted culture was used as inoculum. To confirm that the observed cellulase on starch was not due to carry-over cellulose experiments were conducted by inoculating cellulose-grown culture into broth devoid of carbon source or broth containing glucose or cellobiose. No or little detectable cellulose was observed. On starch the yield of CMCase was higher than that of amylase. Further more, the yield of CMCase per mg substrate utilized was higher on starch than on cellulose (Table 3).

Table 3. Production of cellulose and amylase by *C. tertium* grown on starch or cellulose (at 4 g/l) for 6 days

Substrate	Enzyme	Units/ml	Units/mg substrate utilized
Starch	CMCase	0.05	0.030
	Avicelase	0.11	0.070
	Amylase	0.03	0.017
Cellulose	CMCase	0.08	0.025
	Avicelase	0.06	0.076
	Amylase	ND	–

ND - not detected.

Clostridium tertium ferments various sugars, the purity of this organism was confirmed by microscopical examination. It ferments cellulose even after several transfers on cellobiose. *Clostridium tertium,* in its ability to ferment various carbohydrates resembles *C. thermocellum M7* reported by Lee and Blackburn (1975).

Clostridium tertium when grown in cellulose concentrations greater than 50 g/l, accumulates large amounts of sugars which repress the cellulase activity. Therefore, it is helpful to identify strains of *C. tertium* which are more efficient in bioconversion of cellulose or more adaptable to practical fermentation conditions, thereby increasing the available gene pool for further strain selection and development.

REFERENCES

Avgerinose, G.C., Fang, H.V., Biocio, I. and Wang, D.I.C. (1981). In *Advances in Biotechnology,* Vol.2, Permagon Press, p. 119–124.

Ferguson, T.J. and Mah, R.A. (1983). Isolation and characterization of an H_2-Oxidizing thermophilic methanogen. *Applied and Environmental Microbiology,* **45:** 265–274.

Giuliano, Christine and Khan, A.W. (1984). Conversion of cellulose to sugars by resting cells of a mesophilic anaerobe *Bacteroides cellulosolvens. Biotechnology and Bioengineering.* **27:** 980–983.

Gorden, J., Jiminez, M., Cooney, C.L. and Wang, D.I.C. (1978). AICHE Symp. Ser., 74 (181): 91–97.

Herrero, A.A. and Gomez, R.F. (1980). Development of ethanol tolerance in *Clostridium thermocellum:* Effect of growth temperature. *Applied and Environmental Microbiology.* **40:** 571–577.

Khan, A.W., Lamb, K.A. and Forgie, M.A. (1987) Formation of esterase, xylanase, and xylosidase by cellulolytic anaerobes. *Biomass,* **13:** 135–146.

Lee, B.H., and Blackburn, T.H. (1975). Cellulase production by a thermophilic *Clostridium* species. *Applied Microbiology,* **30:** 346–353.

Lynd, L.R. (1989). Production of ethanol from lignocellulosic materials using thermophilic bacteria. Critical evaluation of potential and review. *Adv. Biochem Engg/Biotechnol.* **38:** 1–52.

Lynd, L.R., Cushman, J.H., Nichols, R.J. and Wyman, C.E. (1991). Fuel ethanol from cellulosic biomass. *Science,* **251:** 1318–1323.

Miller, G.L. (1959). Use of Dinitrosalicylic acid reagent for determination of reducing sugars. *Anal. Chem,* **31:** 426–428.

Vijaya, T. and Dev, D.V. (1997). Studies on factors affecting ethanol production from cellulosic wastes by a mesophilic *Clostridium* sp. *The Journal of the Indian Botanical Society.* 1–3.

Vijaya, T. and Dev, D.V. (1999). Effect of pretreatment of lignocellulosic materials on ethanol and other by-products by a mesophilic *Clostridium* sp. *Journal of Microbial World.* **1(2):** 47-50.

Weimer, P.J. and Zeikus, J.G. (1977). Fermentation of cellulose and cellobiose by *Clostridium thermocellum* in the obsence and presence of *Methanobacterium thermoautotrophicum. Applied and Environmental Microbiology.* **33:** 289–297.

14

Lending in the Biomass Energy Sector

V. Bakthavatsalam

Biomass in its variety of forms is a key source of renewable energy for use as solid, liquid and gaseous fuels. The production and utilization of biomass is a primary need for the energization of rural areas in India and most other developing countries. Different epochs of history have given preference to different forms of energy. During the Industrial Revolution, coal was the most favored form, which brought about farreaching changes in the commercialization of the traditional processes. This was followed by oil, which helped to usher in many significant transformations, particularly in the transportation and industrial sectors. Since the 1970s serious thought began to be given to search for alternative, renewable and nonpolluting sources of energy, such as small, mini and micro hydro, solar and bio-energy. Bioenergy offers very great scope due to a wide spectrum of biomass available under different agro-climatic conditions.

Out of the total solar energy on earth, the plant life utilize about 0.1% annually, resulting in the annual net production of 2×10^{11} tonnes of organic matter which has an energy content of 3×10^{20} J. The total annual energy use, however, is of the order of 3×10^{12} J. One of the natural assets of most developing countries is abundant sunshine. The total solar radiation received over the landmass of India is about 5×10^{10} kWh/Year, with 250–300 days of useful sunshine per year in most parts of the country. The daily average direct radiation at places in the central part of the country is 5–7 kWh/m^2. There is thus a vast scope for harvesting solar energy and improvement in photosynthetic efficiency.

The word 'biomass' is a very comprehensive term comprising all forms of matter derived from the biological activities and present either on the surface of the soil or at different depths of the vast bodies of water—lakes, streams, rivers, seas and oceans. An indicative list of bioresources is presented in Table 1.

Bio-Resources In India

ORGANIC RESIDUES						
Animal Wastes	Forest Wastes	Crop residues/ byproducts	Industrial and Urban Wastes	Firewood Trees/ Shrubs	Carbohydrate Plants	Aquatic Biomass
Cattle dung	Sawdust	Bagasse	Dairy waste	*Acacia auricullformis* (Bengali babul)	*Beta velgiris* (Sugarbeet)	Algae
Poultry excréte	Wood chips	Banana peel/ stem	Distillery effluents	*Eucalyptus*	*Manihot esculenta* (Cassava)	*Eichhornis*
Goat and sheep dropping		Coconut waste	Plastic waste	*A. nilotica* (babul)	*Saccharum officinarum* (Sugarcane)	
		Coffee waste	Textile waste	*Carmaidulensis* (redgum)		
		Corn stover	Urban solid wastes	*A. senegal* (gum arabic)		
		Cotton stalks	Urban liquid wastes	*Eucalyptus tereticornis* (Nilgiri)		
		Ground stalks		*A. tortillis* (Israeli Babul)		
		Groundnut shell		*Hardwickia binata* (Anjan)		
		Jute sticks		*Albizzia lebbak* (Siris)		
		Maize cobs		*Lacunae lleucocephala* (Subabul)		
		Molasses		*Cassia siamea* (Bombay blackwood)		
		Rice husk		Prosopis chilensis (Pardeshi babul)		
		Rice straw		*Casuarina equisetifolia* (Saru)		
				Prospis cinetania (Khejiri)		
				Dalbergia sissoo (shisham)		
				Prospis Juliflora (masquits)		
				Sarbania grandlflora (agathi)		
				Terminalia arjuna (arjun)		

Technologies to convert biomass into energy fall into two categories, biological and thermo-chemical. Biological processes like anaerobic digestion and alcoholic fermentation involve enzymatic breakdown of biomass by microorganism at low pressure and low temperatures. By contrast, termochemical processes use high temperatures to convert biomass into energy by direct combustion, pyrolysis, gasification and liquefaction. An outline of the various biomass energy conversion processes and products is presented in Fig. 1.

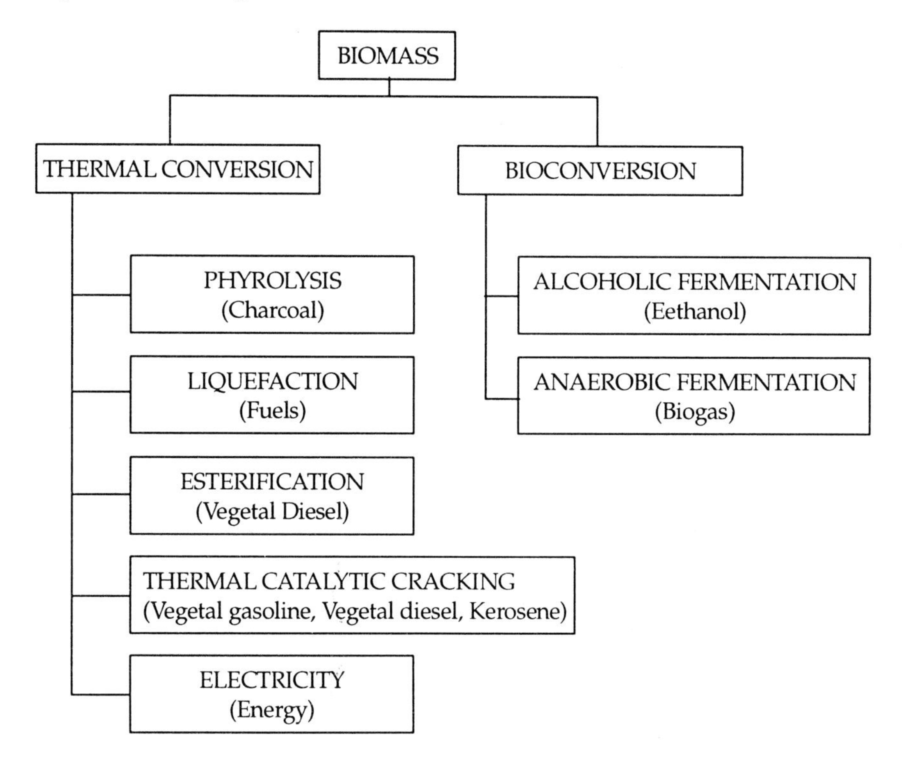

Fig. 1. Biomass conversion processes and products

There has been a general perception that renewable energy technologies and biomass energy can only give small amount in certain locations and therefore their contributions to total energy requirements would be only marginal. Another perception is that these are rather expensive compared to the conventional alternatives. Both perceptions have been unfounded by the momentum and growth in renewable energy technologies and their application in the last few years, especially in India. Significant amounts of electric power, heat, mechanical energy technologies, furthermore the economics of these technologies have also improved considerably, mainly due to recent policy initiatives and incentives package by the Government of India and improved and efficient technologies.

The potential of biomass energy in India is :

Energy Sources	Estimated Potential
Bio-energy	17,000 MW
Bagasse Cogeneration	3500 MW

Renewable Energy Programmes and Bio-energy

In India, the Renewable Energy Programme began with the formation of the Commission for Additional Sources of Energy (CASE) with the responsibility of formulating programmes for the development of NRSE, coordinating and intensifying R&D activities and ensuring the implementation of all Government policies in this regard. Appreciating the importance of developing the NRSE in a larger framework, the Govt. of India created the Department of Non-Conventional Energy Sources (DNES) in September 1982, which later became the Ministry of Non-conventional Energy Sources (MNES). Taking into any account the limitations of the conventional banking approach and in order to accelerate the momentum of development and large-scale utilization of renewable energy sources and primarily for promoting, developing and financing NRSE technologies, the Indian Renewable Energy Development Agency Ltd. (IREDA) was incorporated in March 1987. Within the institutional framework of the IREDA renewable energy technologies including bio-energy have been promoted through research and development, demonstration projects, programmes supported by government subsidies, programmes based on cost recovery supported by IREDA and private sector projects. The physical progress of various subsectors of bio-energy as on March 1999 is given in Table 2.

Table 2. Physical Progress of Biomass Energy Technolgoies in India (Cumulative Achievement up to March 1999)

Sl.No.	Programme	Unit	Achievement
	BIO-ENERGY		
1.	Biomass based co-generation	MW	134
2.	Biomass combustion based power	MW	37
3.	Biomass gasifiers/Stirling engines	MW	32
4.	Family-size biogas plants	Million Nos.	2.7
5.	Improved cook stoves	Million Nos	29.5

The IREDA Experience

Affordable financing is one of the crucial factors inhibiting the usage of renewable energy, especially at the small-user level. IREDA's Mission is to *"be a pioneering, participant friendly and competitive institution for financing and promoting self-sustaining investment in energy generation from renewable sources and energy efficiency for sustainable development".* A major role of IREDA is to provide both renewable energy users, manufacturers and producers credit that initially feature concessional terms but progressively approach commercial market rates as the technology gains wider acceptance. By financing new ventures in renewable energy, IREDA helps create performance track records for NRSE technologies, facilitating their transition from novelty to mainstream status.

The highlights of IREDA's general financing norms are given in Table 3.

Table 3.

Quantum of Assistance	Up to 80 % of the Project cost Up to 90 % of the equipment cost
Rate of Interest	0 % to 15 %
Moratorium	Maximum 3 years
Repayment Period	Maximum 10 years

The details of various debt instruments that IREDA operates in the bio-energy sector are given in Tables 4, 5 and 6. .

Table 4. Project Financing

Sector	Interest Rate	Maximum repayment period including moratorium	Maximum moratorium	Minimum Promoters contribution	Term Loan
Biomass co-generation (including sugar industry)	18.00	10	3	25	Upto 75 % of total Project cost.
Biomass power Generation	10.50	10	3	25	Up to 75 % of total project
1. Municipal waste	16.50	10	3	25	
2. Other than municipal waste					Up to 75 % of total Project
Biomass gasifier for Power generation (Above 500 kW)		5	2	25	Up to 75 % of total Project
biomethanation from industrial effluents		8	2	25	Up to 75 % of total Project
Biomass Briquetting		10	2	25	Up to 75 % of total Project
Biogas plants utilizing animal dung. Human excreta and night soil (20-100 cum/day)	16.50	8	2	20	Up to 80 % of total Project
–Financial intermediary	10.50	8	2	20	
–Direct Users					Up to 80 % of total Project

Table 5. Loans for Manufacturing of Equipment

Sector	Interest Rate	Maximum repayment period including moratorium	Maximum moratorium	Minimum Promoters Contribution	Term loan/ lending norms of IREDA
Biomass briquetting equipment	17.00	7	2	25	Upto 75 % of total project cost.
Biomass gasifiers	17.00	7	2	25	Upto 75 % of total project cost
Vehicles based on alternate fuels such as ethanol/ methanol	17.00	7	2	25	Utpto 75 % of total project cost

Table 6. Equipment Financing

Sector	Interest Rate	Maximum repayment period including moratorium	Maximum moratorium	Minimum Promoters Contribution	Term Loan
Gasifiers (upto 500 KW)	17.50	4	1	20	Upto 80 % of the cost of eligible equipment(s)
Vehicles based on alternate fuels such as ethanol/methanol etc.	17.50	5	1	20	Upto 80 % of the cost of eligible equipment(s)

Within the initial 12 years of operation, till 31st March 1999, IREDA has pledged resources for 1126 Renewable Energy projects pledging resources to the tune of over Rs. 260 crores. Of these, the sector-wise details of IREDA's achievements in the bio-energy technologies are given in Table 7.

Table 7. Financing Biomass Energy Technologies

Sl. No.	Sector	No. of Projects	Installed Capacity
1.	Biomass/Bagasse Cogeneration	19	252.00 MW
2.	Biomass power generation	6	40.75 MW
3.	Bio-methanation	64	953407 m^3/day
4.	Biomass briquetting	31	1450.8 tpd

During the current year and in the forthcoming 9th Five Year Plan also IREDA will be giving continued thrust to biomass energy technologies.

During the past decade, as a result of various Research, Development, Demonstration, Extension, Market Development and Commercialization Programmes, in India and elsewhere, much useful experience has been gained and we are now in a better position to express well founded opinion about various biomass energy technologies and their prospects for the future.

It is clear that bio-energy technologies will play a major role in developing and powering the nation and other similar developing countries, although they may not play a dominant role when compared to their fossil fuel counterparts. With a proper policy environment and with adequate institutional and financing mechanisms in position, the bio-energy technologies will find and occupy their market niches, and constitute an integral part of energy systems of India and rest of the developing world. IREDA invites participation from all concerned in this global renewable energy movement.